Disclaimer

The publisher of this book is by no way associated with the National Institute of Standards and Technology (NIST). The NIST did not publish this book. It was published by 50 page publications under the public domain license.

50 Page Publications.

Book Title: FASTLite: Engineering Tools for Estimating Fire Growth and Smoke Transport (NIST SP 899)

Book Author: Rebecca W. Portier; Richard D. Peacock; Paul A. Reneke;

Book Abstract: FASTLite is a collection of procedures which builds on the core routines of FIREFORM and the computer model CFAST to provide engineering calculations of fire phenomena for the building designer, code official, fire protection engineer and fire safety related practitioner. This manual provides documentation and examples for using FASTLite. It describes how to install the software on a computer and provides a guide for the use of FASTLite using an example.

Citation: NIST SP - 899

Keyword: computer programs; fire growth; smoke transport; computer models; evacuation; fire models; hazard assessment; human behavior; predictive models; toxicity

U.S. DEPARTMENT OF COMMERCE

TECHNOLOGY ADMINISTRATION

National Institute of Standards and Technology

Special Publication 899

FASTLite: Engineering Tools for Estimating Fire Growth and Smoke Transport

Rebecca W. Portier
Richard D. Peacock
Paul A. Reneke

Special Publication 899

FASTLite: Engineering Tools for Estimating Fire Growth and Smoke Transport

Rebecca W. Portier
Richard D. Peacock
Paul A. Reneke

Building and Fire Research Laboratory
Gaithersburg, MD 20899

April 1996

U.S. Department of Commerce
Ronald H. Brown, *Acting Secretary*
Technology Administration
Mary L. Good, *Under Secretary for Technology*
National Institute of Standards and Technology
Arati Prabhakar, *Director*

National Institute of Standards
and Technology
Special Publication 899
Natl. Inst. Stand. Technol.
Spec. Publ. 899
86 pages (April 1996)
CODEN: NSPUE2

U.S. Government Printing Office
Washington: 1996

For sale by the Superintendent
of Documents
U.S. Government Printing Office
Washington, DC 20402-9325

Bibliographic Information

Abstract

FASTLite is a collection of procedures which builds on the core routines of FIREFORM and the computer model CFAST to provide engineering calculations of fire phenomena for the building designer, code official, fire protection engineer and fire-safety related practitioner. This manual provides documentation and examples for using FASTLite. It describes how to install the software on a computer and provides a guide for the use of FASTLite using an example.

Keywords

Computer models; computer programs; evacuation; fire models; fire research; hazard assessment; human behavior; toxicity

Ordering Information

National Institute of Standards
and Technology
Special Publication 899
Natl. Inst. Stand. Technol.
Spec. Publ. 899
86 pages (April 1996)
CODEN: NSPUE2

U.S. Government Printing Office
Washington: 1996

For sale by the Superintendent
of Documents
U.S. Government Printing Office
Washington, DC 20402-9325

DISCLAIMER

The Department of Commerce makes no warranty, expressed or implied, to users of FASTLite and associated computer programs, and accepts no responsibility for its use. Users of FASTLite assume sole responsibility under Federal and State law for determining the appropriateness of its use in any particular application; for any conclusions drawn from the results of its use; and for any actions taken or not taken as a result of analyses performed using these tools.

Users are warned that FASTLite is intended for use only by those competent in the field of fire safety and is intended only to supplement the informed judgment of the qualified user. The software package is a computer model which may or may not have predictive value when applied to a specific set of factual circumstances. Lack of accurate predictions by the model could lead to erroneous conclusions with regard to fire safety. All results should be evaluated by an informed user.

INTENT AND USE

The algorithms, procedures, and computer programs described in this report constitute a methodology for predicting some of the consequences resulting from a specified fire. They have been compiled from the best knowledge and understanding currently available, but have important limitations that must be understood and considered by the user. The program is intended for use by persons competent in the field of fire safety and with some familiarity with personal computers. It is intended as an aid in the fire safety decision-making process.

CONTENTS

1 Introduction 1
    1.1 Getting Started   1
        1.1.1 Learning FASTLite   2
        1.1.2 Hardware and Software Requirements   2
        1.1.3 Installation   3
    1.2 What Is a Graphical User Interface (GUI)?   4
        1.2.1 Using a Mouse or Other Pointing Device   4
        1.2.2 GUI Terminology   5
        1.2.3 GUI shell menus   8
    1.3 Starting FASTLite   8
    1.4 Overview of FASTLite elements   9

2 Creating Input Data for FASTLite 11
    2.1 Creating a New Input File – A Simple Example   11
    2.2 Details of FASTLite Inputs   12
        2.2.1 Fire Scenario Overview Window   12
        2.2.2 Entering a Title for the Input File   14
        2.2.3 Setting Ambient Conditions   14
        2.2.4 Specifying Simulation Time and Spreadsheet Output   14
        2.2.5 Defining Compartments   15
        2.2.6 Defining Connections for Horizontal Flow   16
        2.2.7 Defining Connections for Vertical Flow   18
        2.2.8 Adding Sprinklers and Detectors   19
        2.2.9 Defining the Fire   20
        2.2.10 Specifying Time-Dependent Fire Curves   23
        2.2.11 Tools   24
    2.3 Running this Example   25
    2.4 Recommended Procedure for Defining New Files   26
    2.5 Opening and Saving Input Files   26
        2.5.1 Selecting an Existing Input File   26
        2.5.2 Creating a New Input File   27
        2.5.3 Saving Input File Modifications   28

3 Running the Fire Model 31
    3.1 Starting the Simulation   31
    3.2 While the Simulation Runs   31
    3.3 Handling Events During the Simulation   32
    3.4 Saving the Results of the Simulation   33

4 An Advanced Example 35
    4.1 The Incident   35
    4.2 Computer Analysis   36
    4.3 Entering the Data   37
        4.3.1 Entering Title for the Input File   38
        4.3.2 Defining Ambient Conditions   38

    4.3.3 Specifying Simulation Time and Spreadsheet Output   38
    4.3.4 Modifying Compartment Geometry   38
    4.3.5 Modifying the Vent Connections   39
    4.3.6 Modifying the Fire Definition   39
    4.3.7 Saving Data File Modifications   40
    4.3.8 Running the Simulation   40
  4.4 Results of the Simulation   41

5 Estimation Tools   43
  5.1 Egress Time   43
  5.2 Sprinkler / Detector Activation   47
  5.3 Atrium Smoke Temperature   51
  5.4 Buoyant Gas Head   53
  5.5 Ceiling Jet Temperature   54
  5.6 Ceiling Plume Temperature   56
  5.7 Lateral Flame Spread   58
  5.8 Law's Severity Correlation   60
  5.9 Mass Flow Through a Vent   62
  5.10 Plume Filling Rate   64
  5.11 Radiant Ignition of a Near Fuel   66
  5.12 Smoke Flow Through an Opening   68
  5.13 Thomas's Flashover Correlation   70
  5.14 Ventilation Limit   72

6 Advanced Features   75
  6.1 Changing Display Units   75
  6.2 Input Editor Error Log   76
  6.3 Copy File   76
  6.4 Print File   77
  6.5 View File   77
  6.6 Delete File   78

7 References   81

FASTLite: Engineering Tools For Estimating Fire Growth and Smoke Transport

Rebecca W. Portier, Richard D. Peacock, and Paul A. Reneke

Building and Fire Research Laboratory
National Institute of Standards and Technology

1 Introduction

Simple (algebraic) equations have been a mainstay of engineering calculations for as long as they have existed. With the arrival of modern calculators and then computers, the complexity of these equations has increased, since it was no longer necessary to refer to details such as tables of logarithms to evaluate them. Today we can do fully time-dependent calculations with systems of differential equations on our desktops at the push of a button.

Individual equations that relate to fire phenomena have been around a long time as well, often representing correlations to experimental data and observations. In 1984 Bukowski suggested that a *series* of individual calculations could be used to evaluate a complex, interactive process, i.e., a fire hazard analysis [1]. A broader series of equations applicable to fire growth estimates was also published in 1985 [2]. Nelson [3] extended this concept further with FIREFORM and FPEtool to provide simple models along with engineering calculations in a software package that is widely used for fire safety engineering calculations.

FASTLite (which stands for **F**ire Growth **a**nd **S**moke **T**ransport) is a collection of procedures which builds on the core routines of FIREFORM and a simplified version of the computer model CFAST [4] to provide engineering calculations of fire phenomena for the building designer, code enforcer, fire protection engineer and fire-safety related practitioner. This manual provides documentation and examples for using FASTLite. It describes how to install the software on your computer and provides a guide for the use of FASTLite.

At the outset, it is important to note that FASTLite is intended for use only by those competent in the field of fire safety and is intended only to supplement the informed judgment of the qualified user. The software is intended to provide quantitative estimates of some of the likely consequences of a fire and the underlying models have been subjected to a range of verification tests to assess the accuracy of the calculations. Like any computer calculation however, the quality of the calculated result is directly related to the quality of the inputs provided by the user. The software may or may not have predictive value when applied to a specific set of factual circumstances. Inappropriate use could lead to erroneous conclusions. All results should be evaluated by an informed user.

1.1 Getting Started

FASTLite is intended to be a tool to allow estimation of fire spread in buildings. The software and documentation included on the distribution CD ROM can be installed by following the instructions included with the CD ROM. Although installation of the software is intended to be simple and one can have the software "up and running" quickly, efficient use of all of the features of the software will occur after reading the documentation The user is advised to begin this process by scanning the table of contents to become familiar with the contents of the manual.

1.1.1 Learning FASTLite

Developing appropriate inputs for a realistic simulation of typical fire scenarios can be imposing. While typical values are supplied for much of the information, nearly all of the data can be customized for each specific test case. Details of all of these inputs are included in this document. This section provides a possible strategy to guide one through the documentation.

1. "Getting Started" provides a general description of the software along with a summary of the limitations of the predictive equations used in FASTLite.

2. Beginning on page 4, there is a detailed description of each element of the user interface from using a mouse and keyboard with the software to each type of display used for input of data and presentation of results for the model. For some users, this section will be a review. For others, some study will be required to become familiar with the concepts of a graphical user interface.

3. Section 2, on page 11, begins with a simple example of a fire in a single compartment and provides a step-by step guide to running this example. Reviewing this example will familiarize the user with the basic operation of the software.

4. Section 5, on page 43, describes a number of estimation tools which are included with FASTLite and can be used to provide estimates of individual fire phenomena. While some of these tools require more detail than a couple of input variables, most routines require relatively little effort to obtain an estimate.

5. Finally, refer to section 2 on page 12, for a detailed description of each of the inputs for FASTLite which can be used to customize an individual test case.

1.1.2 Hardware and Software Requirements

- A 386 or later IBM[*] compatible PC with at least 2.5 megabytes of free extended memory. FASTLite does not run on 8086 microcomputers.

- At least a VGA compatible graphics display. The installation program automatically determines video hardware for each installation and sets the software accordingly.

- The mouse driver must be 100% MS-mouse compatible. If problems are experienced, such as the mouse not responding within the user interface, check to be certain the mouse driver was loaded

---

[*] The use of company or trade names within this report is made solely for the purpose of identifying those computer hardware or software products operationally compatible with FASTLite. Such use does not constitute any endorsement of those products by the National Institute of Standards and Technology.

when the computer was turned on. If the driver was loaded and problems persist, update the version of the driver software by contacting the manufacturer of the mouse.

- Do not run memory resident (TSR) programs that were originally developed for 8086 platforms while trying to run FASTLite. Older TSR applications may not take advantage of EMS and XMS memory use and can cause operating system conflicts with the newer technology in graphics displays and interfaces.

FASTLite on a PC looks first for XMS memory, then expanded memory, then unmapped EMS memory. It utilizes the first type that is available and cannot mix types. Thus, if there is expanded memory available, but not enough, the error message "insufficient memory to load EXP file" is displayed.

1.1.3 Installation

An installation program which prompts for the necessary information and copies the necessary files onto a hard disk is included on the distribution CD. FASTLite is a DOS-based program and should be installed directly from the DOS prompt. If necessary, exit MS-WINDOWS or any menuing software prior to installation.

To install FASTLite on a PC computer system, place the CD into the CD ROM drive (typically DOS drive D:) and enter the following DOS command:

```
D:INSTALL
```

If your computer is configured so that the CD ROM drive is assigned a drive letter other than D:, replace the D: in the command above with the appropriate drive letter. Several questions are asked about the computer system and how and where to install the modules. Answer none, some, or all of the questions as appropriate for your specific needs. Follow the directions on the screen closely and provide answers to questions as desired. Usually, the defaults provided by the installation program are sufficient. On several of the screens, prompts are provided to fill in information or change the defaults suggested by the installation module if desired. These screens are described in more detail below.

```
NIST              FASTLite, Version 2.2                    BFRL


      +-------------------------------------------------------+
      |The installation will copy all of the data  and program|
      |files for FASTLite to the hard disk of your selection. |
      |Multiple subdirectories will be created for the program|
      |files and data files.                                  |
      |                                                       |
      |INSTALL will ask you questions about the installation. |
      |Each question has a default answer.  If the default    |
      |answer is acceptable, press the ENTER key in response  |
      |to the question. Otherwise, type the answer or use the |
      |arrow keys to select an appropriate response and then  |
      |press the ENTER key.                                   |
      |                                                       |
      +-------------------------------------------------------+



               [Enter] to continue.  [Esc] to quit.
```

FASTLite is installed in a single subdirectory structure on a hard disk. The first two screens prompt for the drive letter of the hard disk (usually C:), and the name of the main program directory (where the actual software is installed). For most installations, the defaults are sufficient provided the drive selected contains at least 12 megabytes of free space. For custom installations, any drive containing at least 12 megabytes of free space may be selected. Any valid nonexistent directory may be used for the program directory.

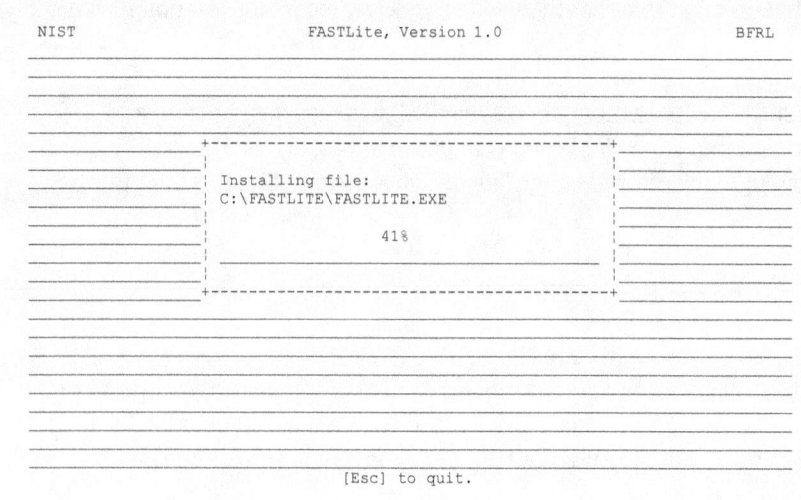

As the files are copied from the installation CD to the hard disk, the installation program shows the progress on the screen.

In order to operate correctly, the installation module may need to create or modify the FILES= statement in the DOS startup file CONFIG.SYS. If you wish to modify the file yourself, you may skip this step. The FILES= statement must specify a minimum of 20 files for the software to operate correctly.

1.2 What Is a Graphical User Interface (GUI)?

The FASTLite shell is a style of interface commonly referred to as a graphical user interface or GUI (pronounced GOO-EEE). Graphical refers to the use of lines, rectangles, colors, and shading to produce a three-dimensional interface appearance on a two-dimensional screen. A familiar GUI for many personal computer users is the MS-WINDOWS interface. The shell supports many of the features familiar to users of MS-WINDOWS applications but is not a MS-WINDOWS application. It is started from the DOS command line.

In order to provide a common base of familiarity, this section will discuss GUI concepts which are common between the application modules in FASTLite.

1.2.1 Using a Mouse or Other Pointing Device

A pointing device such as a mouse or trackball is a hand held device frequently used in GUI applications to communicate selection information to the computer. For this guide, the term "mouse" will be used generically to indicate any appropriate pointing device. This guide references three types of mouse action. **Click** or **single-click** involves positioning the screen pointer on the item of interest and pressing and releasing the left mouse button once. **Double-click** involves positioning the screen pointer on the item of interest and pressing and releasing the left mouse button twice in rapid succession. A **drag** operation enables the user to move windows or items within windows to different positions on the screen. Dragging items involves positioning the screen pointer on the item of interest, pressing the left mouse button, moving the screen pointer to a new position while keeping the button depressed, then releasing the button when the item is positioned in the new location.

1.2.2 GUI Terminology

GUI interfaces are designed to assist the user in running multiple applications similar to an office worker having several project file folders open at one time. Using this parallel, the term **desktop** is used to refer to the full-screen background of a GUI application. A **window** is a rectangular, framed image drawn on the

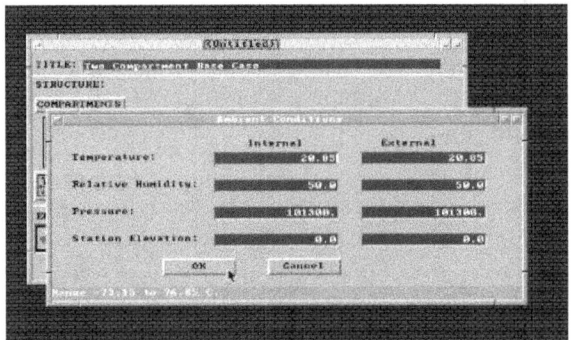

desktop and is used to view carefully selected portions of the larger application. Typically, the top frame, or **title bar**, of the window is used to provide a text title describing the purpose of that window. This clearly distinguishes one window on the desktop from other windows displayed at the same time.

When several windows are concurrently displayed, a problem could arise as to which window the user intends to apply keyboard entries. This issue is resolved in GUI applications through the concept of the **current** or **focus** window. The title bar display characteristics of the focus window are unique in comparison to all other currently displayed windows. The title bar of the focus window has a solid background while other windows have a shaded or color muted background.

 In the top left corner of some windows, a horizontal bar is displayed inside a small square. Any window with this horizontal bar displayed can be removed from the screen, referred to as **closing the window**, by positioning the mouse pointer on this line and double-clicking.

Two types of windows are used within the FASTLite interface. The most common is referred to as a **dialog** or **modal window**. Modal windows are typically used for user input and can be identified by the row of buttons at the bottom of the window to allow the user to accept or reject the values shown in the window. The Ambient Conditions window, above, is an example of a modal window. When a modal window is displayed, the user cannot interact with the application through any other window or menu. The user must complete input for the current window or reject current input values in order to interact with other GUI windows or menus. Current input is accepted by clicking on the **OK** item. Current input values are rejected by clicking on the **Cancel** item or pressing the **Esc** key. Modal windows cannot be moved or sized.

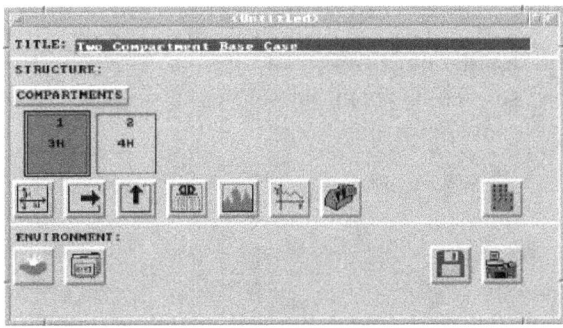

The **nonmodal window** or **overview window** is used as a background window in the GUI shell to summarize the current state of input. In FASTLite, the most frequently used nonmodal window is the fire scenario overview window. Nonmodal windows can be moved by the user. To move a nonmodal window, position the mouse pointer on the title bar and drag the window to a new position using the technique described in section

1.2.1. To make a nonmodal window the current or focus window, single-click on the title bar for that window.

The elements within GUI windows used to display information and request input from the user are referred to as **widgets**. Widgets within FASTLite have special display characteristics, responses, and associated functionality dependent on the widget type. Not all widgets support selection by the user. Refer to the detailed descriptions of each widget type in this section to determine selectability, functionality, and display characteristics. To move from one selectable widget to widgets below or to the right of the current widget, press the Tab key, or click on the desired widget using the mouse. To move to widgets above or to the left of the current widget, hold the Shift key down and press the Tab key, or click on the widget using the mouse. The following widgets are used at different times within the FASTLite interface:

Labels: Labels are used to prompt for the type of information requested in an associated edit widget. Labels are easily identified by the use of black text with no outline frame. Labels cannot be selected or modified by the user.

Edit widgets: Edit widgets are visually identified by white text displayed on a blue background. Edit widgets along with edit lists are the only widgets for which keyboard entry by the user is displayed to the screen. For integer and floating point edits, a message is displayed at the bottom of the window indicating the allowable range of input values for the current edit widget. If an entry is invalid, the range message is replaced with an error message displayed on a red background when the user attempts to select another widget. Corrections must be made to an invalid entry before the interface allows selection of another widget.

Text Buttons: Text buttons are used to initiate user selected operations such as accepting or rejecting input or displaying additional windows for related input. The associated functionality is activated by a single mouse click on the button, or by pressing the Tab key until the screen pointer is positioned on the button then pressing Enter. Two special text buttons are available on most dialog windows. The *OK* button indicates to the application that entries in the edit widgets for this window are to be accepted and applied to the input specification. This acceptance pertains to the current use of the input editor, and does not affect the associated file stored on disk. To modify the disk file contents the user must save the input file following the procedure discussed in section 2.5.3. The *Cancel* button indicates to the application that entries in the edit widgets for this window are to be ignored, and the values which existed prior to use of this window are to be retained.

Pull-down widgets: Pull-downs are used to provide the user with a complete list of all possible entries for a prompt field when the list is short and pre-determined by the application.

Graphic icons: Graphic icons use pictures to represent the functionality of a button. The associated functionality is activated by a single mouse click on the button, or by pressing the Tab key until the screen pointer is positioned on the button then pressing Enter.

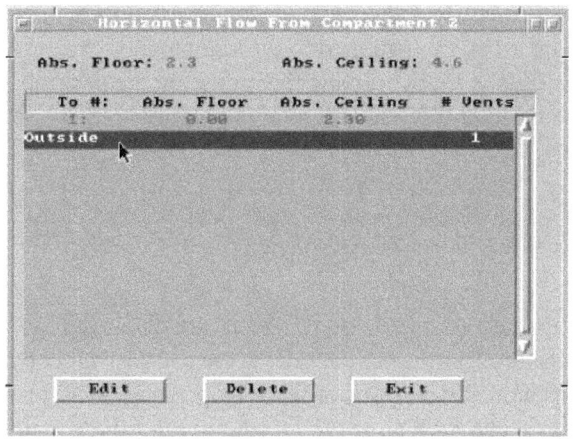

Selection lists: Selection lists are used to provide a complete list of all possible options when that number potentially exceeds the size of a window and is dependent on previous selections. Only one entry may be selected from the list. The current entry selected is highlighted by a blue background with white text while the remaining available entries are indicated by a cyan background with white text. A selection is made by *double-clicking* with the mouse, by highlighting the desired selection with a single-click and pressing Enter, or by pressing the ↑ or ↓ keys to highlight the selection then pressing Enter.

Scroll bars: At times, the current window size is insufficient to display all available information, *e.g.*, a selection list contains more lines than are available for display on the screen. To accommodate the additional information, a vertical scroll bar is attached to the display window when the additional information is to be displayed in a top to bottom list. When the additional information is displayed in a left to right manner, a horizontal scroll bar is attached to the display window in order to access additional, previously undisplayed information. The displayed window performs a function similar to the use of a magnifying glass for viewing images on paper. Through the use of the appropriate direction arrows, the user is able to move the viewer up and down, or left and right, across the associated image. To view entries preceding, or above, the currently displayed entries in the window, press the up arrow icon (▲) on the vertical scroll bar. Press the down arrow icon (▼) to view entries below the currently displayed entries. If the horizontal scroll bar is available, press the left arrow icon (◄) or right arrow icon (►) to pan the current viewer left or right across the associated image. Each time the user clicks on a direction arrow, the viewer scrolls one line or character in the corresponding direction. The rectangular box on the scroll bar indicates the current position within the full image to the user. To scroll the equivalent of one window in a direction, click on the appropriate scroll bar adjacent to the current position box in the direction scrolling is desired. If the user wishes to scroll down one window, click immediately below the current position box on the vertical scroll bar. To scroll right one window, click to the right of the current position box on the horizontal scroll bar.

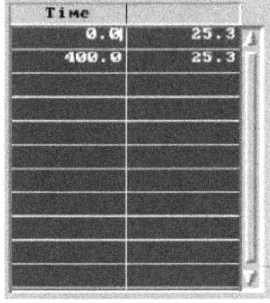

Edit lists: Edit lists are spreadsheets with display characteristics similar to the edit widgets. Background for each cell is blue with white text, and range messages are displayed at the bottom of the window. The range displayed is dependent on the context of the cell or block of the spreadsheet in which the text cursor is displayed. Range checks are activated in a manner similar to the edit widgets prohibiting the user from exiting a cell until the entry is valid. Movement between cells in an edit list requires pressing either the Shift-→ or Shift-← to move between columns from any position in a cell. Press the ↑ or ↓ to move between rows.

Cursors: Two types of cursors are used in a GUI interface. In addition to the blinking cursor (▮), or "I-beam," familiar to users of text-oriented interfaces, an **arrow pointer** (➤) is used to indicate the current mouse position. This becomes important when the user enters input or presses the Enter key from the keyboard since the widget responding to the input is *always* the focus widget. If the focus widget is an edit widget or an edit list, the I-beam is displayed inside the widget or cell to indicate where keyboard entry will be applied. If the focus widget is a text or graphics button, keyboard entry is applied to the widget only when the arrow pointer is displayed in that widget. A new focus widget is set by moving the mouse on the desktop then single-clicking, or by pressing the Tab key on the keyboard until the arrow pointer is pointing to the desired widget.

1.2.3 GUI shell menus

Menus provide lists of available applications or functions for selection by the user. Menus are displayed by clicking on the corresponding icon, or for some special menus, on a portion of the GUI screen. Once a menu is displayed, functions are selected by clicking on the item of interest with the mouse, or by pressing the ↑ or ↓ keys until the desired item is highlighted then pressing Enter. The FASTLite GUI shell provides three main menus of interest: the **desktop menu**, the **file menu**, and the **toolbox menu**.

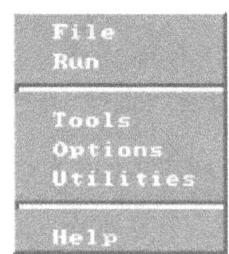

The **desktop menu** is used to select other modules within the FASTLite suite of applications. The name indicates that the desktop menu can be activated by clicking on the GUI desktop, the screen background. An alternate means for displaying the desktop menu involves selecting the desktop icon from the fire scenario overview window within the GUI shell. A summary of the applications associated with each of the desktop menu items is presented in section 1.4 below.

The **file menu** provides the user with a way to save input information in the currently selected file or in another filename specified by the user. Selection of the **save** option overwrites the current file while selection of the **save as** option prompts the user to specify a new filename. The file menu can be activated by clicking on the title bar of one of the fire scenario overview window within the GUI shell. An alternate means for displaying the file menu involves selecting the disk icon from the fire scenario overview window within the GUI shell.

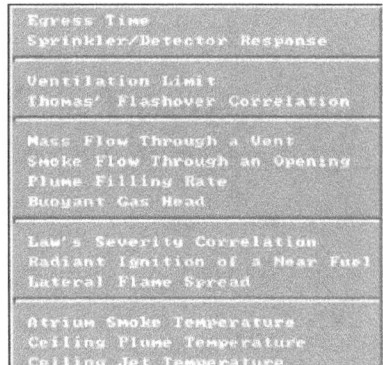

The **toolbox menu** provides access to several quick calculation routines. Details of each routine are discussed in section 5 below. The toolbox menu can be activated by displaying the desktop menu then selecting the tools option. An alternate means for displaying the toolbox menu involves selecting the toolbox icon from the fire scenario overview window.

1.3 Starting FASTLite

To start FASTLite, change to the directory which contains the program and run it by changing to the directory where the software was installed and executing the program. The following commands are typical:

```
cd \fastlite
fastlite
```

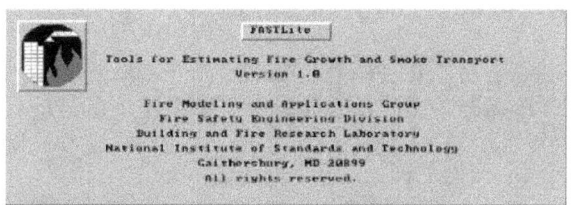

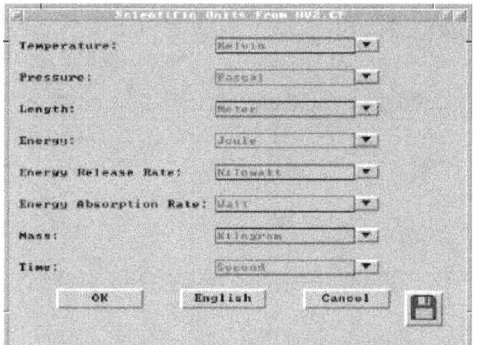

and are typed from the DOS command prompt (pressing the Enter key after each command). The FASTLite logo window is displayed. Press Enter to display the desktop menu from which other modules of the application suite can be selected. An overview of the FASTLite elements is presented in the next section. For detailed discussions, refer to the individual sections of this guide corresponding to desktop menu items of interest.

Two windows are automatically displayed the first time the program is executed:

Measurement Units: Desired measurement units are selected with the measurement units windows.. Any combination of units may be used as desired. For the example cases in this manual, common metric units are used. Units used for temperature, pressure, length, energy, energy release rate, energy absorption rate, mass and time are: Celsius, pascal, meter, joule, kilowatt, watt, kilogram and second.

To modify the settings for any of the base measurements, click on the pull-down icon to the right of the measurement. A pull-down menu is displayed, listing the available measurement units. Select the display units desired by clicking on the corresponding menu entry. For example, temperature can be displayed in Kelvin, Celsius, Rankine, or Fahrenheit. Select *OK* to close the measurement units

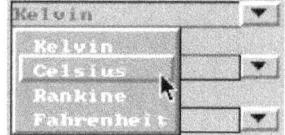

window. Measurement units may be changed at any time by selecting *Options* from the desktop menu, then selecting *User Specified Units* to display the measurement units window. For further details on customizing the units see section 6.1.

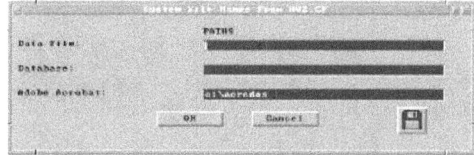

System Files: The location of program data files, databases, and documentation reader are specified in the system files window. Normally, the entries in this window can be left at their default values. If FASTLite or the Adobe Acrobat Reader™ was installed in a directory other than the default provided by the installation program, change the entries in this window to reflect the proper location. Select *OK* to close the system files window. The system file locations may be changed at any time by selecting *Options* from the desktop menu, then selecting *System Files* to display the system files window. For further details, see section 6.1.

1.4 Overview of FASTLite elements

FASTLite is composed of several interdependent modules which provide the different analyses desired by the individual user. The desktop menu discussed previously groups these modules according to the type of work and typical order in which a user might choose to execute the applications. This section provides a brief discussion of the applications and groupings with references to the appropriate section in this guide for detailed information.

File: Specify details of the structure and fire scenario to be analyzed. For an example, refer to section 2.

Run: Run FIREFORM 3.2: Run the original DOS-based version of FIREFORM [3] included with FPEtool version 3.2. Most of the calculations included in FIREFORM have been duplicated within FASTLite and are available from the tools menu below.

Tools: Quick calculation tools which can be used to determine individual characteristics of the fire, *e.g.*, occurrence of flashover or mass flow through a vent. For details, refer to section 5.

Options: Customize the installed FASTLite suite of applications by specifying measurement display units, and licensing information. For details, refer to section 6.

Utilities: View, copy, and print files. For details, refer to section 6.

Help: Provides an overview of each graphic icon on the main FASTLite window and provides information on the version number and telephone support contacts for FASTLite.

2 Creating Input Data for FASTLite

FASTLite is an interactive, user-friendly program used to generate input data files for and run the FASTLite model. As such, it is difficult to describe all the functions of the program in a reference guide. Rather, it is best learned through use. This section will describe the types of information entered on each of the windows of FASTLite. Users unfamiliar with the concepts of a Graphical User Interface (GUI) are encouraged to review these concepts in the Getting Started section of this guide before continuing with the Input Data section. Familiarity with GUI terminology and the use of a mouse is assumed throughout the remainder of this section.

2.1 Creating a New Input File – A Simple Example

To aid the first-time user in becoming familiar with the operation of FASTLite, this section provides a simple example of the use of FASTLite and includes detailed instructions of the use of the software for the example. For this example, we will simulate a fire in a two room structure. Doors will connect the two rooms of the structure and connect the structure with the outdoors. The fire source will be a growing fire typical of combustibles found in single-family residences.

To start FASTLite, change to the directory which contains the program and run it by changing to the directory where the software was installed and executing the program. The following commands are typical:

```
cd \fastlite
fastlite
```

from the DOS command prompt (pressing the Enter key after each command). If the software was installed to a different directory, substitute an appropriate directory name for the `cd \fastlite` commend above.

If this is the first time the program has been run since being installed, desired measurement units must be first selected. See section 1.3 for details on setting initial measurement units. After the FASTLite logo appears, the main desktop menu is displayed. Using the mouse, click on *File* then *New* to begin the definition of a new fire scenario. Initially, there are two steps to defining a new fire scenario to be modeled – selection of the geometry of the structure (the number of compartments and connections between the compartments) and a description of the fire.

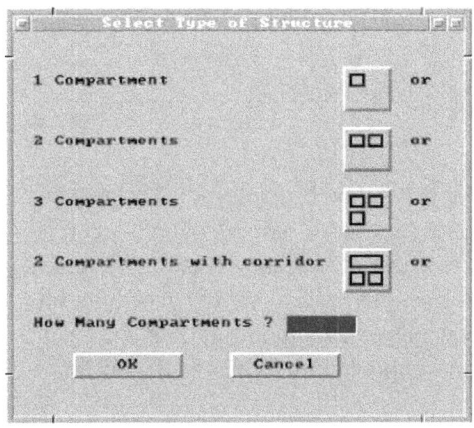

Once *New* is selected, the structure selection window is displayed. From this window, the number of compartments and the arrangement of the compartments is defined. Once an initial selection has been made, the size of the compartments and any connections between compartments can be customized to fit the details of a specific scenario. For this example, choose the "2 Compartment" example by using the mouse to select the 2 compartment icon. Note that the selection highlights the icon with a bold black border. This selection defines two compartments 2.4 m wide by 3.6 m deep by 2.4 m high with doorways opening to the outdoors and connecting the compartments. Click on *OK* to accept the selection and close the window.

Following selection of the structure, the fire must be specified. For a wide range of fires, the fire growth can be accurately represented with a power law relation of the form

$$\dot{Q} \propto \alpha \, t^2 \qquad (1)$$

where Q is the heat release rate of the fire, α is the fire intensity coefficient, and t is time [5]. A set of specific T-squared fires labeled slow, medium, and fast, with fire intensity coefficients (α) such that the fires reached 1055 kW (1000 BTU/s) in 600, 300, and 150 seconds, respectively were proposed for design of fire detection systems [6]. Later, these specific growth curves and a fourth called "Ultra-fast" [7] which reaches 1055 kW in 75 seconds, gained favor in general fire protection applications.

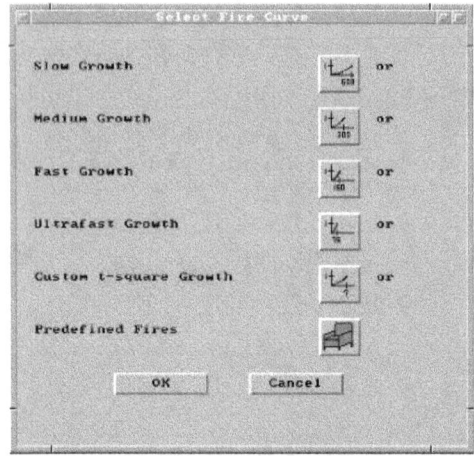

For this example, a medium T-squared growth rate fire will be specified. In a mixed collection of fuels selecting the medium curve is appropriate as long as there is no especially flammable item present. In a manner similar to the selection of the "2 Compartment" structure geometry above, select the medium growth rate fire by using the mouse to select the "Medium Growth" icon. Note that the selection highlights the icon with a bold black border. Click on *OK* to accept the selection and close the window. A new window is then displayed prompting for specific times to define the fire curve. For this example, the default values for the medium growth rate fire of 300 s and 900 s will be used. Click on *OK* to accept these defaults and close the window.

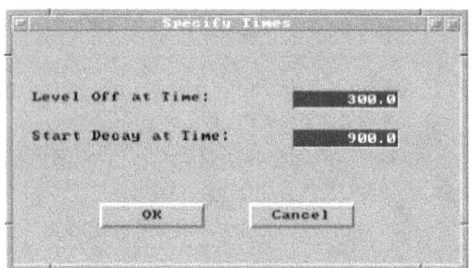

This completes the description of the example. Execution of the example case will continue in section 2.3. The next section describes the elements of FASTLite and provides a reference for all of the inputs to the software.

2.2 Details of FASTLite Inputs

2.2.1 Fire Scenario Overview Window

Once an input file has been created or opened, the fire scenario overview window is displayed. This window enables the user to see at a glance the major characteristics of the input file prior to running the model. It is possible to quickly determine the number of compartments (# of boxes drawn in structure section of window), the fire compartment (red box), and compartments with some type of sprinkler or detector (cyan or light blue boxes). Since the current FASTLite model is concerned only with flow between compartments and not with the physical adjacencies of those compartments within the structure, no implications should be drawn from the adjacency of boxes displayed in this window to physical adjacencies of the corresponding structural compartments.

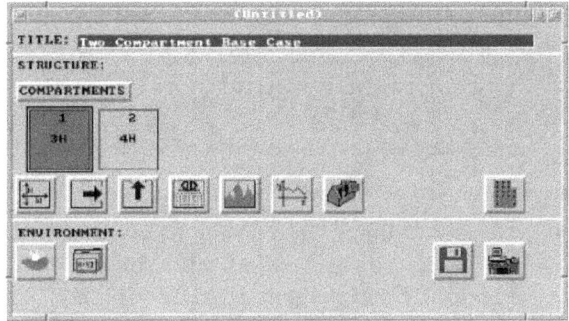

The fire scenario overview window is divided into three sections. The upper section provides the title for the input file. The second section pertains to the definition of compartments, flows between the compartments, placement of sprinkler/detectors, and the definition of fire curves. The final section of this overview window provides background information required to run the fire model such as the ambient conditions, filenames, and output times. Details for entering information into each of these sections are presented below.

In the first section of the overview window, each input file can be assigned a unique text description for identification purposes.

The structure overview or middle section of the window shows the number of compartments currently defined. For each compartment, text within the box indicates number of horizontal flow vents, *e.g.*, doors and windows (a number followed by H), and number of vertical flow vents, *e.g.*, vents in the ceiling or floor of the compartment (a number followed by V). The color of each compartment provides an indicator of the fire compartment (red) and/or compartments which have detection/suppression devices (cyan).

Prior to activating any of the graphics icons in the structure section of the overview window, a compartment must first be selected to determine how the activated icon will function. Click on compartment #1 of the input file, then press the geometry icon. Dimensions and surface materials for compartment #1 are displayed. Click on Cancel to close the geometry window, click on compartment #2 of the input file and press the geometry icon. The dimension and surface material details now reflect specifications for compartment #2. In a similar manner, experiment with the context changes in information displayed by pressing each of the graphics icons for different compartments. In the case of the time curve icon, the menu of possible curve definitions is dependent on the selected compartment along with other information provided for that compartment. If the compartment has a fire, the time curve icon allows entry of several different fire curves for that compartment. If the fire does not exist for the currently selected compartment, the time curve icon is grayed out and disabled. Each of the graphics icons in this middle section of the window can be matched to an icon image in section 2.5.2 in order to determine the related function and type of information requested.

The third section of the fire scenario overview window provides background information for the fire model run. Ambient conditions such as temperature and relative humidity are specified by pressing the ambient conditions icon. Length of simulation time along with print and spreadsheet files for the model run can be entered using the filenames icon.

Two final graphics icons are placed in the bottom right corner of the window. These icons are:

Write to file: save and save as. Select save to save changes to the current file or save as to save changes to a new file leaving the currently selected file unchanged. Refer to section 2.5.3.

 **Desktop:** Provides access to the desktop menu which allows the user to select other modules within FASTLite; perform file utilities such as copy, print, or delete, and view files such as the current input file or the input editor error file generated when errors are found in opening an existing input file. Refer to section 1.4 for discussion of each of these modules.

2.2.2 Entering a Title for the Input File

Before beginning to define the structure, it is typically best to start by assigning a title or description to the input file. The title is not used by the model, but is intended as an aid for the user in cataloguing the various fire scenario input files. Tab to the title field or move the mouse cursor to this field and click with the left mouse button. Enter the text desired. All titles maintain upper and lower case sensitivity as entered by the user.

2.2.3 Setting Ambient Conditions

 To add or edit the internal and external ambient temperature, pressure, and station elevation along with information on external wind, Tab or move the mouse to the ambient conditions icon in the environment section of the fire scenario overview window, and click or press Enter.

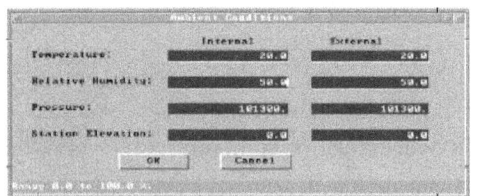 Note the display of range and measurement units at the bottom of the window as each edit widget is made the focus widget. To customize the units see the Advanced Features section 6.1. Internal refers to conditions inside the structure while external indicates conditions outside the structure. Fields to be entered by the user along with default values are indicated below.

Internal Temperature: Initial temperature inside the structure. Default value 20 °C.

External Temperature: Initial temperature outside the structure. Default value 20 °C.

Relative Humidity: Initial relative humidity inside and outside the structure. Default values are 50% for the internal and external.

Pressure: Default values for inside and outside are standard atmospheric pressure at sea level, 10.13 kPa.

Station Elevation: The height where the pressure measurements were taken. This is the reference datum for calculating the density of the atmosphere as well as the temperature and pressure inside and outside of the structure as a function of height. Default value is 0, which is sea level.

2.2.4 Specifying Simulation Time and Spreadsheet Output

 To specify the simulation time, display time, or output to spreadsheet file with output interval, Tab or move the mouse to the filename icon in the environment section of the fire scenario overview window, and click or press Enter.

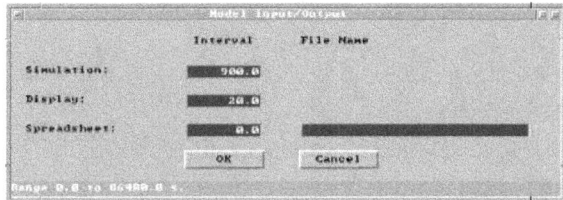

Note the display of range and measurement units at the bottom of the window as each time edit widget is made the focus widget. To customize the units see the Advanced Features section 6.1.

Simulation: The length of time over which the simulation takes place. This is a required input which should be entered even if all other fields on the window are left unchanged. If the *File, New* option was used, a default value equal to the time the decay started plus the growth interval is provided.

Display: The time interval between each graphical display of the output for runtime graphics. This output time pertains to the selection of the Graph button on the simulation status window and specifies how often the graph is updated. A default dependent upon the simulation time is provided.

Spreadsheet: The spreadsheet file stores a subset of the output of the model at specified intervals in a comma-delimited alphanumeric format which can be read by most spreadsheet software. This is designed to be imported into a spreadsheet for further analysis or graphing of the results of the simulation. If a file name is entered, a default time interval dependent upon the simulation time is provided.

2.2.5 Defining Compartments

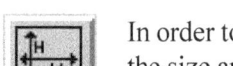

In order to model a fire scenario, the user must portray the geometry of the structure in terms of the size and elevation of every compartment in the structure. Thermophysical properties of the enclosing surfaces must also be specified by selecting surface materials in order to accurately model the transfer of heat through the surfaces. The maximum number of compartments is three. Connections between compartments are defined later in sections 2.2.6 and 2.2.7.

Two types of "add compartment" functionality are available. Compartments can be appended to the end of the currently displayed graphics list or inserted between existing compartments. If append is used, and two compartments currently exist, the new compartment would be compartment 3. If insert was selected with two compartments existing, and the current compartment is compartment 2, the new compartment would be numbered 2 and the current compartment would become 3. Compartments can be added to the structure by clicking on the compartment text button or by tabbing to the open area below the compartment button and pressing *INS* or *Ctrl-I* to insert, and *Ctrl-A* to append.

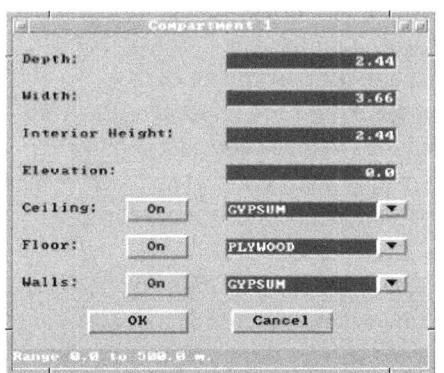

After selecting the type of add desired, the geometry window is displayed. Note the display of range and measurement units at the bottom of the window as each edit widget is made the focus widget. To customize the units see the Advanced Features section 6.1.

Depth: Depth of the compartment as measured forward from the left, rear corner of the compartment.

Width: Width of the compartment as measured across from the left, rear corner of the compartment.

Interior Height: Height of the compartment.

Elevation: Height of the floor of the compartment with respect to the station elevation specified in the ambient condition window.

Ceiling, Floor, Walls: The *On/Off* button turns conduction for the associated surface on and off within this compartment. When the text "Off" is displayed on the button, the surface of the button is grayed out, or stippled. The adjacent text field for the thermal material is also stippled. The text field specifies the thermal material to use for conduction when the *On* button has been pressed. This material name can be entered by the user or selected from the database. To select a thermophysical material from the database, enter as much of the material database key word as is known, click on the pull-down icon, scroll the database until the desired material is viewed, and select by double-clicking, or highlight and press Enter. If the material key word is entered by the user, it must be a valid entry in the thermal properties database before the compartment can be generated.

Deleting Existing Compartments

To delete an existing compartment, select the compartment in the structure graphics list on the fire scenario overview window. Press *Alt-D* or the *Delete* key to delete the compartment.

2.2.6 Defining Connections for Horizontal Flow

Horizontal flow connections may include doors between compartments or to the outdoors as well as windows in the compartments. These specifications do **not** correspond to physically connecting the walls between specified compartments. Lack of an opening prevents flow. Horizontal flow connections may also be used to account for leakage between compartments or to the outdoors. Opening and closing of these connections is not handled by the fire model at this time.

To specify a horizontal flow opening, select the corresponding compartment in the structure graphics list on the fire scenario overview window. The determination of which of a pair of compartments should be selected as the first compartment is not important since defining flow connections from compartment 1 to 2 immediately implies connections from compartment 2 to 1. Flow connections do not indicate direction of the flow, but simply enable the flow to occur. Direction of flow is determined by the characteristics of the

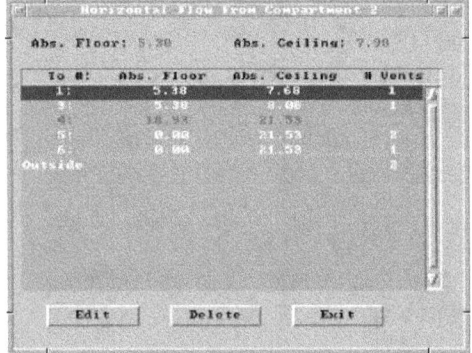

physical phenomena such as pressure and temperature at the time the model is run. The true significance of selection of a from compartment is reflected in the positioning of the connection. Once a compartment has been selected as the from compartment, it will become the reference compartment for specification of the position of the connection. Tab or move the mouse to the horizontal flow icon, and click or press Enter.

All compartments with currently defined elevation and height of compartment are displayed in the selection list. Horizontal connections can only be created between compartments that physically overlap in elevation at some point. Those compartments for which such a connection is possible are highlighted in white in the selection list. Compartments which cannot be connected because of this physical constraint have been grayed out. To complete the connection, click on the compartment in the selection list to which flow is desired and press the Edit button, or double-click on the selection with the mouse.

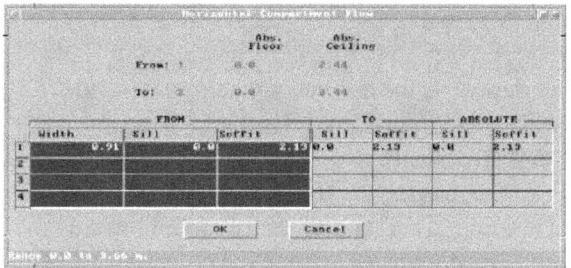

Note the display of range and measurement units at the bottom of the window as each cell in the edit list widget is made the focus widget. To customize the units see the Advanced Features section 6.1.

It is possible to define a total of four (4) horizontal flow connections between any pair of compartments using the spreadsheet. In order to complete each flow definition, the size and location of the connection with respect to the floor of the compartment must be entered. There is no provision for the specification of the placement of a vent along the length of the wall. Vent numbers are automatically defined when the file is saved in order to uniquely identify each of the connection pairs. The vent number assigned corresponds to the row of the spreadsheet in which the entries are made. The user is free to enter vent information on any of the four rows provided. To move to a cell either click on that cell using the mouse, or use Shift-→ and Shift-← to move right and left, and ↑ and ↓ to change rows.

Width: The width of the opening.

Soffit: Position of the top of the opening above the floor of the compartment selected as the first compartment from the graphics list. This compartment is displayed as the from compartment on this window with corresponding elevation and height.

Sill: Sill height is the height of the bottom of the opening above the floor of the compartment selected as the first compartment from the graphics list. This compartment is displayed as the from compartment on this window with corresponding elevation and height.

Deleting existing connections: To delete all existing connections between a pair of compartments, from the selection list, highlight the compartment to which all connections are to be deleted, and press the Delete button. To delete individual connections between a pair of compartments, select the compartment to which the connection is to be deleted, and press the Edit button. Move to the spreadsheet row representing the vent to be deleted, and press Alt-d to delete the entry or Alt-e to erase the entry. Deleting the entry will move all entries below the current row up one row resulting in a renumbering of remaining vents. Erasing the entry will leave all other rows in current positions and will retain current vent number assignments.

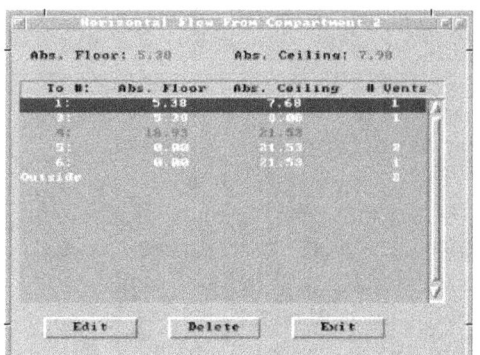

2.2.7 Defining Connections for Vertical Flow

Vertical flow connections include any vertical flow openings, such as holes between compartments or to the outdoors. These specifications do **not** correspond to physically connecting the ceiling and floor between specified compartments. Lack of an opening prevents flow. Vertical flow connections may also be used to account for leakage between compartments or to the outdoors. Opening and closing of these connections is not handled by the fire model at this time.

Vertical flow connections are specified by selecting one compartment as the current compartment in the structure graphics list on the fire scenario overview window. The determination of which of a pair of compartments should be selected as the first compartment is not important since defining flow connections from compartment 1 to 2 immediately implies connections from compartment 2 to 1. Flow connections do not indicate direction of the flow, but simply enable the flow to occur. Direction of flow is determined by the characteristics of the physical phenomena such as pressure and temperature at the time the model is run. The significance of selection of first compartment is reflected in the specification of the position of the connection. Once a compartment has been selected as the first compartment, it will become the reference compartment for positioning of the connection.

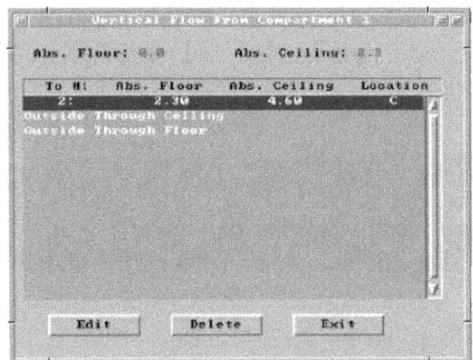

Tab or move the mouse to the vertical flow icon, and click or press Enter. All compartments with currently defined elevation and height of compartment are displayed in a selection list. Vertical connections can only be created between compartments that could be physically stacked based on specified floor and ceiling elevations for the compartments. Some overlap between the absolute floor height of one compartment and the absolute ceiling height of another compartment is allowed. However, whether the compartments are stacked or overlap somewhat, the ceiling/floor absolute elevations must be within 0.01 m of each other. Those compartments for which such a connection is possible are highlighted in white in the selection list. Compartments which cannot be connected because of this physical constraint have been grayed out. To complete the connection, click on the compartment to which flow is desired and press the Edit button, or double-click on the selection with the mouse.

Note the display of range and measurement units at the bottom of the window as each edit widget is made the focus widget. To customize the units see the Advanced Features section 6.1.

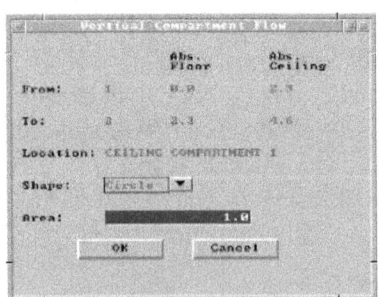

It is possible to define only one vertical flow connection between any pair of internal compartments. However, two connections may be specified from a compartment to the outdoors, one through the ceiling and the other through the floor. In order to complete each definition, the shape and area of the opening must be entered. Once a shape has been selected, the maximum possible area for a vent with that shape is calculated by the input editor. This value will be the upper bound allowed for the area of the opening. There is no provision for the specification of the placement of the vent along the width or breadth of the ceiling. The position of the vent in the floor or ceiling of a compartment is automatically calculated by the input editor based upon the specified elevations for the pair of compartments.

Shape: Circle or square.

Area: The effective area of the opening. For a hole, it would be the actual opening. For a diffuser, the effective area is somewhat less than the geometrical size of the opening.

Deleting existing connections: To delete an existing connection between a pair of compartments, from the selection list, highlight the compartment to which the connection is to be deleted, and press the Delete button.

2.2.8 Adding Sprinklers and Detectors

Sprinklers and detectors are both considered detection devices by the FASTLite model and are handled using the same input window. Detection is based upon heat transfer to the detector. Fire suppression by a user-specified water spray begins once the associated detection device is activated. A maximum of 20 sprinklers or detectors can be included for any input file and model run. These can be in one compartment or scattered throughout the structure. Sprinklers and detectors are added by selecting the compartment into which the device is to be placed from the structure graphics list in the fire scenario overview window. Tab or move the mouse to the sprinkler icon, and click or press Enter.

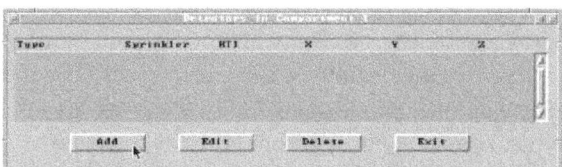

To change the specification for an existing sprinkler or detector, highlight the entry and press the Edit button, or double-click with the mouse. To add a new sprinkler or detector, press the Add button.

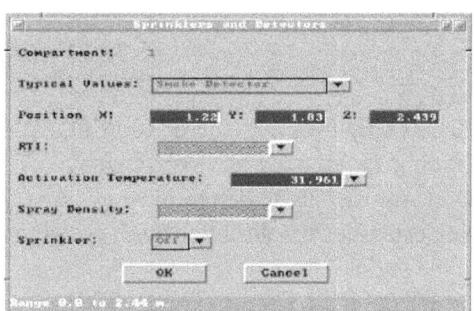

Note the display of range and measurement units at the bottom of the window as each edit widget is made the focus widget. To customize the units see the Advanced Features section 6.1.

In order to complete the definition, the type of detection device, position of the detector within the compartment, and the type of sprinkler must be specified. If the sprinkler is turned on, the detection device is considered a sprinkler. For sprinklers, the response time index (RTI), activation temperature, and spray density must also be entered. Default values are provided for residential and commercial sprinklers.

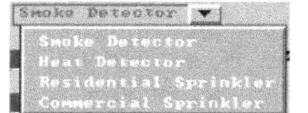

Typical Values: Several sets of default values for typical detection / suppression devices can be chosen. The choice of smoke detector, heat detector, residential sprinkler, or commercial sprinkler effects the default RTI, activation temperature, and sprinkler characteristics below. By default, the device is placed in the center of the compartment at ceiling level.

X Position: Position of the device as a distance from the rear wall of the compartment. See diagram.

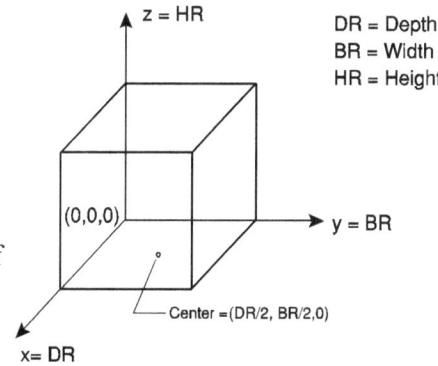

19

Y Position: Position of the device as a distance from the left wall of the compartment. See diagram.

Z Position: Height of the device above the floor of the compartment. See diagram.

RTI: The RTI (Response Time Index) quantifies how rapidly the detector link temperature rises in response to immersion in a hot ceiling jet. For residential sprinklers, a default RTI of 50 $(m \cdot s)^{\frac{1}{2}}$ is assumed. For commercial sprinklers, a default RTI of 100 $(m \cdot s)^{\frac{1}{2}}$ is assumed. RTI is ignored for smoke detectors. Values for typical devices, chosen from NFPA 13 [8], can be selected with the pull-down list (▼) to the right of the *RTI* edit widget. These typical values are representative of often-used designs and are converted into appropriate units for input to the model calculation and entered directly into the *RTI* edit widget.

Activation Temperature: The temperature at or above which the detector link activates. For residential sprinklers, an activation temperature of 57 °C (135 °F) is assumed. For commercial sprinklers, an activation temperature of 74 °C (165 °F) is assumed. Smoke detectors are simulated by an activation temperature of 11 °C above ambient. Values for typical devices can be selected with the pull-down list (▼) to the right of the *Activation Temperature* edit widget. These typical values are representative of often-used designs and are converted into appropriate units for input to the model calculation and entered directly into the *Activation Temperature* edit widget.

Spray Density: The amount of water dispersed by a water spray-sprinkler. The units for spray density are length/time. These units are derived by dividing the volumetric rate of water flow by the area protected by the water spray. The spray density may be measured by collecting water in a pan located within the spray area and recording the rate-of-rise in the water level. Values for typical devices, chosen from NFPA 13, can be selected with the pull-down list (▼) to the right of the *Spray Density* edit widget. These typical values are representative of often-used designs and are converted into appropriate units for input to the model calculation and entered directly into the *Spray Density* edit widget.

Sprinkler: If turned off, the device is handled as a heat or smoke detector only – values entered for RTI, trigger value, and spray density are ignored. The suppression calculation is based upon an experimental correlation by Evans [9], and depends upon the RTI, trigger value, and spray density to determine the behavior of the sprinkler. Two cautions should be observed when using estimates of sprinkler suppression within the model: 1) the first sprinkler activated controls the effect of the sprinkler on the heat release rate of the fire. Subsequent sprinklers which may activate have no additional effect on the fire simulation. 2) The fire suppression algorithm assumes the effect of the sprinkler is solely to reduce the heat release rate of the fire. Any effects of the sprinkler spray on gas temperatures or mixing within the compartment are ignored. Since the detection of heat detectors is based upon a ceiling jet temperature and velocity, the user must take these effects into account when analyzing the actuation of detection devices after sprinkler activation.

2.2.9 Defining the Fire

To define a fire, the compartment of fire origin must be selected and chemical properties entered. Time-dependent combustion properties are entered in section 2.2.10 below. Only one fire is permitted by FASTLite.

To position the fire in a compartment, select that compartment as the current compartment in the structure graphics list in the fire scenario overview window. Tab or move the mouse to the fire icon, and click or

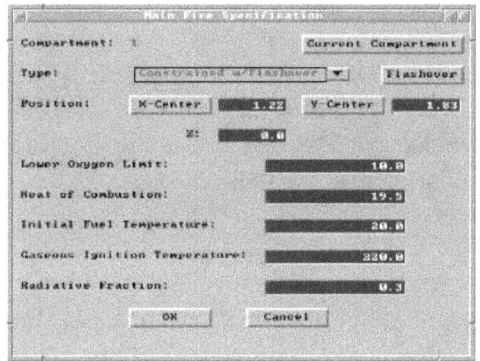

press Enter. Note the display of range and measurement units at the bottom of the window as each edit widget is made the focus widget. To customize the units see the Advanced Features section 6.1.

Type: Unconstrained, constrained, or constrained with flashover. The burning rate for a constrained fire is limited by the available oxygen entrained into the plume. The default is constrained with flashover. If an unconstrained type is selected, only the toxic combustion products can be specified for species. All other species curves are available for a constrained fire type. If constrained with flashover is selected, an additional button is available to defined the post-flashover burning characteristics. To turn the fire off, but retain any associated fire characteristics for future use, select *OFF*.

X Position: Position of the fire as a distance from the rear wall of the compartment. See diagram.

Y Position: Position of the fire as a distance from the left wall of the compartment. See diagram.

Z Position: Height of the fire above the floor. See diagram.

Lower Oxygen Limit: The limit on the ratio of oxygen to other gases in the system below which a flame will not burn. This is applicable only to constrained fires. The default value is 10.

Heat of Combustion: Heat of combustion of the fuel. A default value for wood of 19.5 MJ/kg is assumed. Any value entered directly for the heat of combustion is considered a constant value

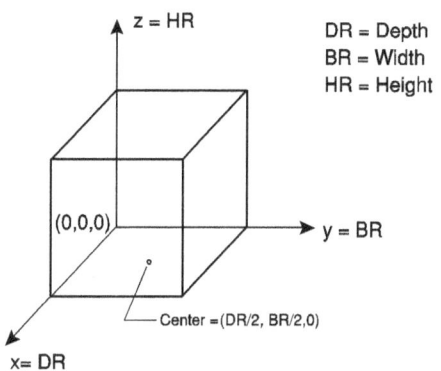

throughout the simulation. If a constant value is desired, fire specification is completed by entering either the pyrolysis rate or the heat release rate curve in section 2.2.10. The remaining undefined variable, pyrolysis rate or heat release rate, is then automatically calculated. If time dependent values are available for the mass loss rate and rate of heat release, enter these values in section 2.2.10 and the model will calculate the corresponding time history for the heat of combustion.

Initial Fuel Temperature: Typically, the initial fuel temperature is the same as the ambient temperature specified on the ambient conditions window. This may not be true if significant conductive or radiative heat transfer affects the fuel.

Gaseous Ignition Temperature: Minimum temperature for ignition of the fuel as it flows from a compartment through a vent into another compartment. If omitted, the default is arbitrarily set to the initial fuel temperature plus 200 °C.

Radiative Fraction: The fraction of heat released by the fire that goes into radiation. A default value of 0.3 is assumed [10]. For other fuels, the work or Tewarson [11], McCaffrey [12], or Koseki [13] is available for reference. These place the typical range for the radiative fraction at a maximum of about 0.6.

Additionally, the fire specification window has several buttons which serve purposes unique to this window.

Current Compartment button allows the user to select from the structure graphics list a compartment which previously did not contain the fire, click on the fire icon, press the *Current Compartment* button, and the fire is relocated to that compartment. All specifications for the fire remain unchanged except for the fire location. The fire position is recalculated so that the relative position in the original compartment is maintained in the new compartment.

X-Center button allows the user to have the X position automatically calculated using previously specified dimensions for the compartment. Pressing this button indicates to the model that the fire should remain in the center of the X position regardless of changes to the compartment dimensions or relocation of the fire to other compartments. Once the user manually enters the corresponding X Position field, however, the X center functionality is terminated.

Y-Center button allows the user to have the Y position automatically calculated using previously specified dimensions for the compartment. Pressing this button indicates to the model that the fire should remain in the center of the Y position regardless of changes to the compartment dimensions or relocation of the fire to other compartments. Once the user manually enters the corresponding Y Position field, however, the Y center functionality is terminated.

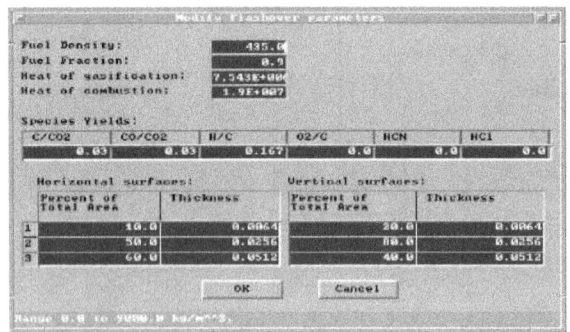

Flashover: Clicking the *Flashover* button displays the flashover screen. The flashover screen has a number for entries of information needed to continue the simulation in flashover. Default values are provided for all fields.

Fuel density: Average density of the fuel burning during post-flashover burning

Fuel fraction: the fraction of the total material that will pyrolyze and lead to combustion.

Heat of gasification: the amount of energy necessary to pyrolyze a unit mass of the fuel – that is turn it from a solid to a gas.

Heat of combustion: the amount of energy a given mass of the fuel gives off when it burns.

Combustion chemistry: The next row of entries defines the combustion chemistry for the fuel. *C/CO2* field is the ratio of mass of carbon produced for smoke to the mass of carbon dioxide produced in combustion of the fuel. *CO/CO2* is the ratio of mass of CO produced to the mass of CO2 produced in combustion of the fuel. *H/C* is the ratio of the mass of hydrogen to carbon in the fuel. *O2/C* is the mass of oxygen liberated to the atmosphere by pyrolization to the mass of carbon in the fuel. *HCN* and *HCl* are the ratio of the mass of HCN and HCl produced to the mass of CO2 in combustion.

Available fuel: These tables are used to put in the amount of fuel available in the room. One table is for fuel with its surface oriented horizontally; the other is for fuel with its surface oriented vertically. For each orientation, up to three thicknesses of fuel can be specified. The amount of fuel is given as a percentage of the total room surface with the same orientation. For example, for horizontal fuel, this is the percentage of the total wall surface area. The area in a table can add up to more than 100% to allow for the inclusion of surface area of furniture, partitions, or other material in the compartment. When all the inputs are satisfactory clicking on the *OK* button continues the run.

2.2.10 Specifying Time-Dependent Fire Curves

Time-dependent curves specify the course of the fire over time for use with the FASTLite fire model. To define a time-dependent curve, the user must first select the fire compartment from the structure graphics list in the fire scenario overview window. If the current compartment is not the fire compartment, indicated by part or all of the current compartment box painted red, time curves specifying combustion characteristics of the fire are not available for input. If an unconstrained fire type is specified, only the toxic combustion products are available. No other species curves may be specified for the unconstrained fire type. If the fire does not exist for the current compartment, no time curve input options are available.

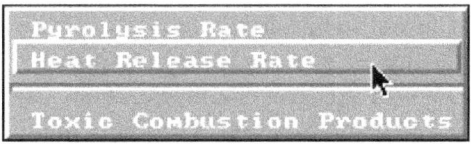

Specifying Time-Dependent Combustion Properties: Select the fire compartment from the structure graphics list in the fire scenario overview window. Tab or move the mouse to the time curve icon, and click or press Enter. Select the desired curve, e.g., *Heat Release Rate*, from the menu.

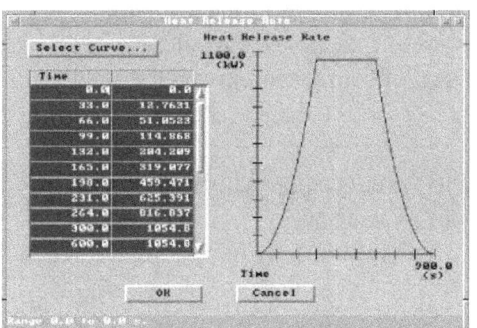

Note the display of range and measurement units at the bottom of the window as each cell in the edit list widget is made the focus widget. To customize the units see the Advanced Features section 6.1. If no other time-dependent curves have been entered previously for this input file, no entries will be displayed in the spreadsheet. To enter time curves for the first time, Tab to the spreadsheet or click on the first cell using the mouse. Enter the actual time at which each fire property is evaluated. Use the vertical scroll bar to enter additional entries beyond those initially displayed. Refer to the GUI Terminology section 1.2.2 of this reference guide for an explanation of the use of scroll bars in the GUI interface.

Movement between time cells is handled using the ↑ and ↓ keys . To insert a new time entry between two existing rows, press *Alt-I*. To delete a row, press *Alt-D*. Deleting a row moves all rows following the deleted row up one row. If the entry is to be erased, but all other rows are to remain in current positions, press *Alt-E* to erase the row. Time entry value ranges are constrained by the entries in surrounding time cells so that time entries within a column of the spreadsheet are ever increasing values. The maximum value allowed is 86400 seconds. Once time entries have been completed for one curve, these time points are available for all other curves. Consequently, only the y-axis values will need to be entered for subsequent curves.

Once the time points have been specified in the first column of the curve spreadsheet, select the top cell of the right column to begin entering the combustion property curve. Keyboard movement from the left column to the right column can be handled by pressing Shift-→ and then the ↑ key until the first cell is reached. Specify the values for the combustion property according to details discussed below. As values are entered in the right column, the fire curve corresponding to time and y-axis entries is plotted on the graph to the right of the spreadsheet.

Pyrolysis Rate: The rate at which fuel is pyrolyzed at times corresponding to each point of the specified fire. Note concerns below regarding the over specification of the fire curves.

Heat Release Rate: The heat release rate of the specified fire. Note concerns below regarding the over specification of the fire curves.

Over specifying Fire Curves: Since the heat of combustion, heat release rate, and pyrolysis rate are related properties, the fire curve can be over specified. If each of the three parameters, heat of combustion from the fire specification window, heat release rate curve, and pyrolysis rate have been specified, the fire is over specified. The input editor accounts for this by using the two most recently entered to calculate the third parameter. This allows for two typical scenarios depending on whether the user desires to use a constant value heat of combustion or a heat of combustion curve. If the user desires to use a constant value heat of combustion, this value should first be entered in the fire specification window. Either the pyrolysis rate or heat release rate curve is then entered. If the pyrolysis rate curve is entered, the heat release rate is automatically calculated by multiplying each entry in the pyrolysis rate curve by the constant value heat of combustion. If the heat release rate is entered, the pyrolysis rate curve is calculated by dividing each entry of the heat release rate curve by the constant value heat of combustion. For the user desiring a heat of combustion curve rather than a constant value, the user should enter the heat release rate and pyrolysis rate curves separately. The model will calculate the appropriate heat of combustion curve prior to execution. One caution regarding this approach. If a user desiring a constant value heat of combustion saves the input file and returns later, the input editor views the heat release rate and pyrolysis rate curves as the last two properties entered. If the user then makes modifications to entries in one of the curves, the other curve will not be automatically calculated. The user must make a change to the heat of combustion in order to get the second curve recalculated.

Toxic Combustion Products: Kilogram of "toxic" combustion products produced per kilogram of fuel pyrolyzed. This is a fraction and is applicable to whatever scientific unit is selected.

2.2.11 Tools

Quick estimation tools are available through the input editor to calculate values such as flashover conditions or mass flow through a vent or to provide estimates for input parameters for FASTLite simulations. Select the compartment to be modeled from the structure graphics list in the fire scenario overview window. Tab or move the mouse to the tools icon, and click or press Enter. The FIREFORM menu is displayed.

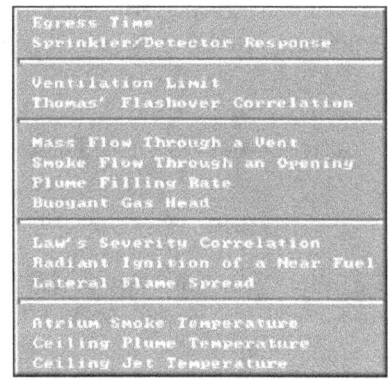

Select the desired estimation tool by clicking on that entry in the displayed menu. An input window for the selected estimation routine is displayed.

Default values based on previously entered specifications for the selected compartment and fire scenario are provided for each appropriate edit widget. Make the appropriate modifications for default values and enter all values for which no default is provided. Refer to the appropriate subsections of section 5 below for additional details for each estimation calculation.

2.3 Running this Example

Once the description of the fire scenario is complete, the fire scenario overview window is displayed. This window enables the user to see at a glance the major characteristics of the scenario prior to running the simulation. It is possible to quickly determine the number of compartments (# of boxes drawn in structure section of window), the fire compartment (red box), and compartments with some type of sprinkler or detector (cyan or light blue boxes).

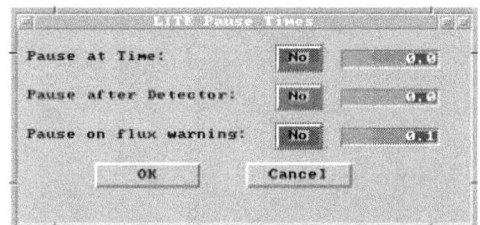

To run the simulation, use the mouse to select the *Run Simulation* icon. After selecting the Run Simulation icon, a window is displayed prompting for times at which the simulation is to pause. These times can be relative to the beginning of the simulation or to the operation of a detection device. For this example, Click on *OK* to indicate no pauses and begin the simulation.

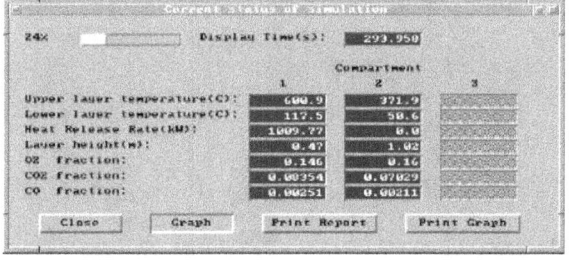

The results of the simulation are displayed in the simulation status window. As the calculation proceeds, the display is updated. The current simulation time is displayed along with calculated values for temperatures, layer interface height, and major gas species. Graphs of temperature, layer interface height, and heat release rate can also be displayed by clicking on the *Graph* button.

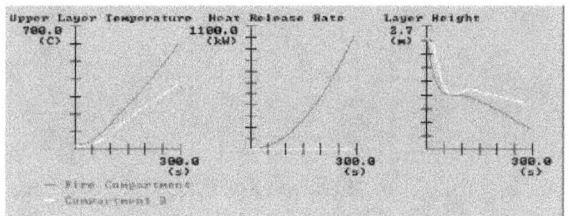

For this example, flashover is reached in approximately 400 s. At this point, the simulation is paused enabling the user to modify the fire description to include the burning of additional combustibles in the compartment. Additionally, the sizes of vents may be changed. For this example, simply click on the *Stop* button to end the simulation. The *Print Report* (to print a summary of the input data and model calculations) and *Print Graph* (to print the three graphs to printers which support the HPGL graphics format) buttons can be used to produce printed documentation of the simulation. For this example, select *Close* to return to the fire scenario overview window.

This completes the example. Additional details on running a scenario are included in section 3. To exit FASTLite, use the mouse to select the desktop icon and display the desktop menu Select the *File* options, then select *Quit*. Since a new simulation has been created and not saved to disk, a window is displayed to allow the option of saving the file prior to exiting the program. Since this example can easily be recreated following the instructions above and is already included as EXMEDIUM.DAT, click on *No* to exit the program without saving a data file to disk.

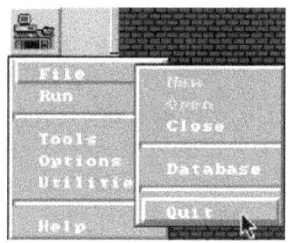

2.4 Recommended Procedure for Defining New Files

Once the number of compartments and fire type have been specified for the new file, the fire scenario overview window is displayed. The user is free to enter information in any order desired with the constraint that compartments should be specified first. As an aid in generating an input file for the first time user, the following sequence is suggested:

- Enter a title for the input file.

- Enter the ambient conditions.

- Specify output files and intervals for acquiring information during the fire model run.

- **SAVE THE FILE**

- Modify compartment geometries for all compartments.

- **SAVE THE FILE**

- Modify and / or define horizontal and vertical connections.

- Position sprinklers and detectors in appropriate compartments.

- **SAVE THE FILE**

- Position fire.

- Modify detailed time curves for the fire including species production, heat release rate, and pyrolysis rate.

- **SAVE THE FILE**

2.5 Opening and Saving Input Files

2.5.1 Selecting an Existing Input File

It is recommended that the new user run through the simple example provided above to gain familiarity with the layout of the overview windows in the input editor. Once the user is familiar with the layout, new input files can be quickly generated following the guidelines in section 2.5.2.

In order to view an existing FASTLite input file, select the *File* option from the desktop menu in the FASTLite GUI shell, then select *Open*. A file specification window is displayed.

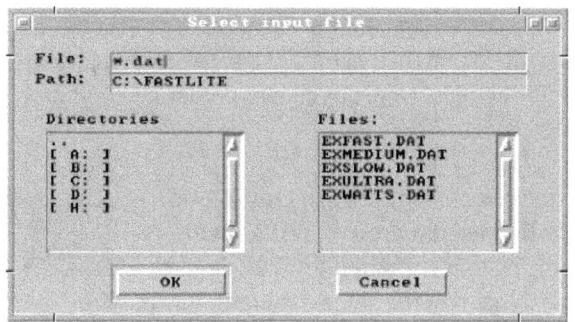

The file specification window consists of four key elements: the filename edit widget, the directory edit widget, the directory selection list, and the filename selection list. To search directories other than the current directory displayed in the directory edit widget, enter the new directory name in the directory edit widget and press Enter. Directories may also be selected by double-clicking on the drive or directory in the directory selection list. If the filename is known, enter it in the filename edit widget and press Enter. Do not include the drive or directory in the filename. Drive and directory specifications are handled only through the directory edit widget and the directory selection list widget. If a partial filename is known, enter the known characters along with * to indicate those parts of the filename for which characters are not known. For example, enter *t*.dat* and press Enter to select data file beginning with the letter *t*. All files matching the specified file name are displayed in the filename selection list. Use the mouse to double-click on the filename desired in the selection list, or highlight the name and click on the *OK* button.

2.5.2 Creating a New Input File

Using the mouse, click on *File* and *New* to begin the definition of a new fire scenario. Initially, there are two steps to defining a new fire scenario to be modeled – selection of the geometry of the structure (the number of compartments and connections between the compartments) and a description of the fire.

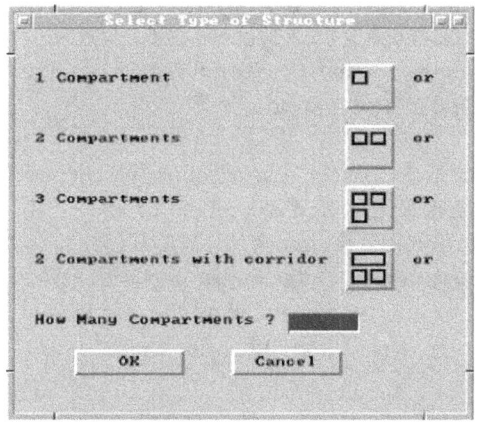

Once *New* is selected, the structure selection window is displayed. From this window, the number of compartments and the arrangement of the compartments is defined. Once an initial selection has been made, the size of the compartments and any connections between compartments can be customized to fit the details of a specific scenario. Use the mouse to select one of the compartment icons (1 compartment, 2 compartments, 3 compartments, or 2 compartments with a corridor). Note that the selection highlights the icon with a bold black border. Each of these four selections creates a structure with an appropriate number of compartments, doors between the compartments and leakage to the outside. Alternatively, by selecting "How many compartments?" the use can create a one, two, or three room structure without defining any ventilation. This later option is useful for creating customized structures for specific know scenarios. Click on *OK* to close the window.

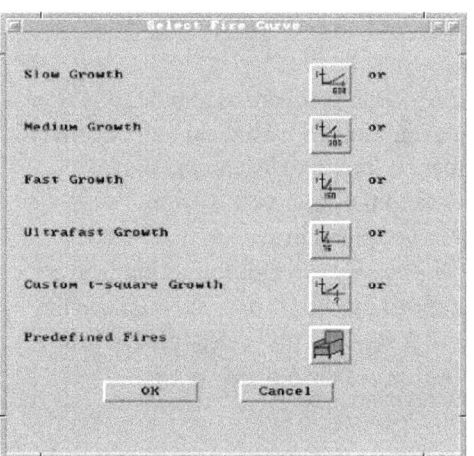

Following selection of the structure, the fire is specified. For a wide range of fires, the fire growth can be accurately represented with a power law relation of the form

$$\dot{Q} \propto \alpha \, t^2 \qquad (2)$$

27

where Q is the heat release rate of the fire, α is the fire intensity coefficient, and t is time. A set of specific T-squared fires labeled slow, medium, fast, and ultra-fast fires with fire intensity coefficients (α) such that the fires reached 1055 kW (1000 BTU/s) in 600, 300, 150, and 75 seconds, respectively are predefined within FASTLite. In addition, a custom growth rate can also be selected.

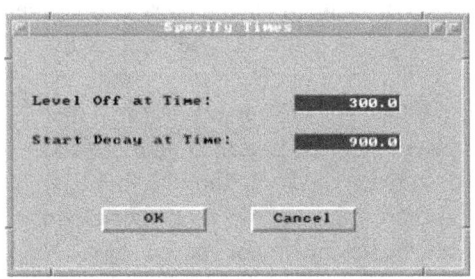

Use the mouse to select one of the fire growth rate icons (slow growth, medium growth, fast growth, ultra-fast growth, custom t-squared growth, or predefined fires). Note that the selection highlights the icon with a bold black border. Click on *OK* to close the window. The Specify Times window is displayed. This window takes two forms.

If one of the standard fire growth rate curves is selected (slow, medium, fast, or ultra-fast), a window is displayed to input a time to the end of the growth phase of the fire (a "Level Off at Time" input) and of the steady burning phase of the fire (a "Start Decay at Time" input) can be defined. If the custom t-squared growth curve is selected, the window also includes the beginning time for fire growth, the time to reach 1 MW and the time at the end of the decay phase of the fire.

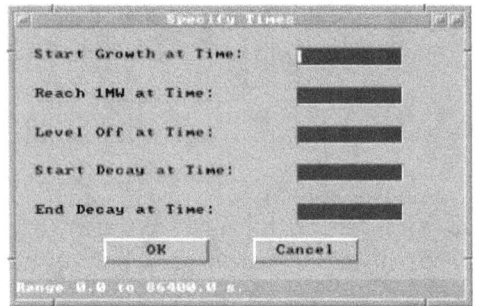

If desired, the default times can be changed by selecting one of the edit widgets and using the keyboard to change the number to any desired value. Click on *OK* to accept the chosen values and close the window.

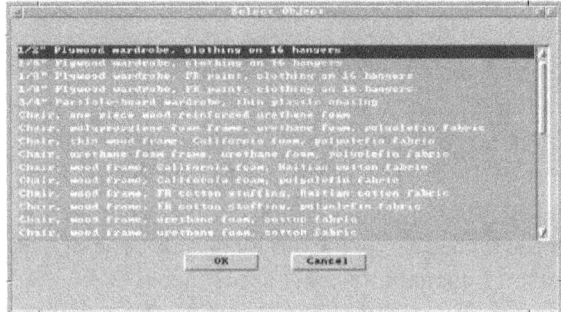

If the predefined fire is selected, a selection list window is display with a description of a number of fire scenarios taken from the literature [5], [14], [15]. Select one of the entries with the mouse and click on *OK* to accept the chosen entry and close the window. This completes the description of the fire scenario.

2.5.3 Saving Input File Modifications

In order to reuse input specifications from one use of the FASTLite software to another, the user must request the input editor to store the information on the computer's hard disk. All information for the current session of the input editor is grouped together on the hard disk and assigned a unique name referred to as the filename for retrieval purposes in the future. It is strongly recommended that input file specifications be frequently saved to disk as outlined in section 2.4. Two options are available for saving input to disk depending on whether the previous set of information is to be retained or replaced. If an existing input file was opened and modifications made, the user may replace the previous specifications with the new specifications by clicking on the diskette icon and selecting the *Save* option. If the previous set of specifications may be needed in the future, select the *Save As* option.

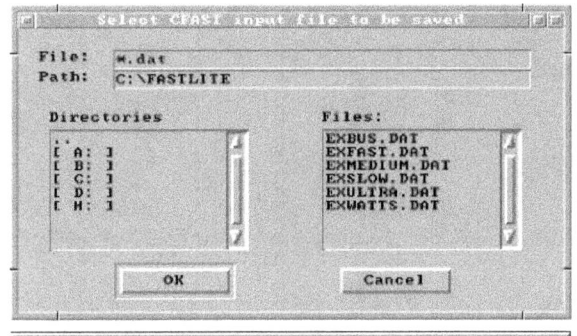

Enter a unique name to be associated with the current set of input specifications. If the filename entered already exists, a warning is displayed. Select *Yes* to overwrite the existing file, *No* to return to the file name prompt window, or *Cancel* to cancel the current save as operation.

Should the user forget to save the current set of specifications prior to exiting the FASTLite shell, a warning box is displayed. Select *Yes* to save the file, *No* to exit without saving the file, or *Cancel* to return to the GUI shell.

3 Running the Fire Model

3.1 Starting the Simulation

 When the description of a fire scenario is completed or an existing file is opened, the fire scenario overview window is displayed. At this point the simulation can be started. To begin the simulation, use the mouse to click on the *Run Simulation* icon.

Before beginning the simulation, a window is displayed to allow the user the option to pause the simulation. The default is to have the simulation run to completion. To activate a pause condition, click on the button associated with the condition to switch it from *No* to *Yes*. Enter the time desired in the corresponding edit field. None, one, or both conditions can be set. Once the pause conditions are set as desired click on *OK* to begin the simulation. Selecting *Cancel* will end the simulation and return to the fire scenario overview window.

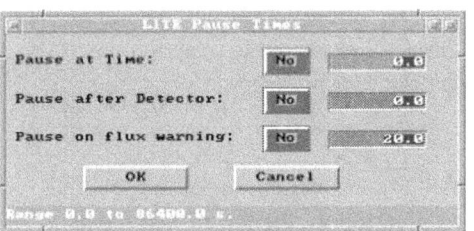

Pause at Time The simulation can be paused at a specific user defined time. This allows existing vents to be opened or closed at some point during the simulation. An example of using this feature is given in section 3.

Pause After Detector The simulation can also be paused at a specified time relative to detector activation. This feature allows for the simulation of automatic doors that close when a detector activates or some other action in response to a detector. If more than one detector is included in a simulation, the pause time applies to all detectors.

Pause at Heat Flux Warning The simulation can also be paused at a user specified heat flux at the center of the floor of the fire compartment. This can be used as another indicator of flashover (values in the range of the default of 20 kW/m^2 are typically used to represent impending flashover) or as an indicator of pain or burns to persons exposed to the fire. For example, a heat flux of 84 kW/m^2 for a duration of 17 s is included in the NFPA standard for protection apparel for firefighters.

3.2 While the Simulation Runs

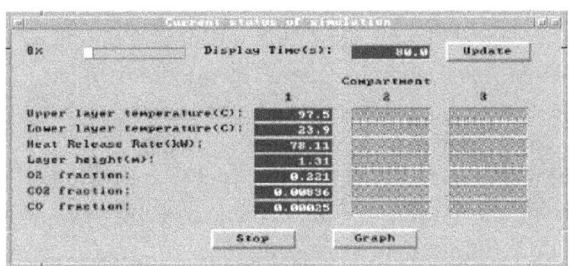

Once the simulation starts, the simulation status window is displayed. The simulation time and other information are automatically updated at intervals specified by the user (in the output file window covered in section 2.3.5). For each compartment the upper layer temperature, lower layer temperature, total heat release rate, fraction of O_2, fraction CO_2 and fraction of CO are listed in a column.

Update Click on the *Update* button to display the current conditions on the simulation status window and the graph window if selected.

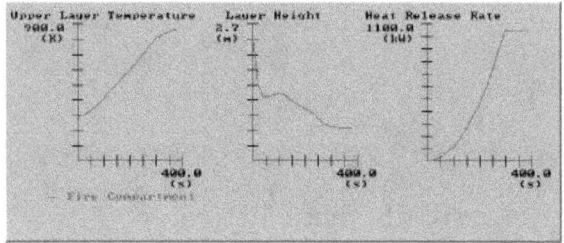

Graph The graph button controls the display of a window showing graphs of the upper layer temperature, the calculated HRR, and the upper layer depth. Once the graph window is displayed clicking on the graph button closes the graph window. In the lower left corner of the graph window is the legend for all three graphs. The fire room is always the red line so that it is easy to identify. When the update button is clicked on the graphs are also updated.

Stop The stop button does as the name implies. It terminates the simulation closes the status window and the graph window if it is open and returns to the fire scenario window.

3.3 Handling Events During the Simulation

Several events can cause the simulation to pause. The first three are in response to inputs from the pause window at the start of the simulation. Finally, the simulation may be paused when flashover conditions are reached – when the upper layer in the fire room reaches 600 °C.

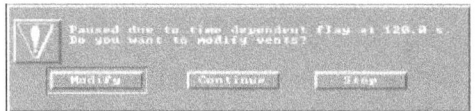

At either the time specified to pause the model, the specified time relative to detector activation, or the time when the heat flux at floor level exceeds a user specified value, a window is displayed with notification of the reason for the pause and offering three options. From left to right are *Modify*, *Continue*, and *Stop*. *Modify* allows the user to change the size of any predefined vent – see below for details. *Continue* returns to the simulation to continue running. *Stop* terminates the simulation and returns to the fire scenario window.

Modify Vents The *Modify* button displays a window listing all the vents available. Only vents that already exist can be changed, new ones cannot be added. A vent can be selected to be modified either one of two ways. First the mouse cursor can be positioned over the vent and selected. Then clicking on the *modify* button

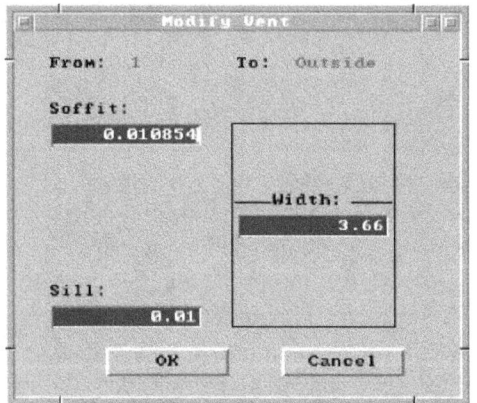

goes to the modify window. The other method is to simply double click on the desired vent in the list. The *continue* button starts the simulation back.

The window for modifying a particular vent in the simulation has three fields for data. The top one gives the current height of the soffit of the vent relative to the floor. The middle field has the width of the vent and the bottom has the height of the sill. All three fields can be edited to change that particular characteristic of the vent. *Ok* returns to the modify vent window saving any new inputs. The *cancel* button returns without changing the values the simulation is using.

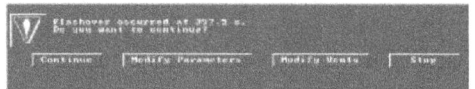

Finally, the simulation may be paused when flashover conditions are reached – when the upper layer in the fire room reaches 600 °C. When this happens, a window is displayed informing that flashover has occurred and allows the user to continue the simulation or to stop. *Continue* allows the simulation to proceed with the user defined fire. *Modify Parameters* allows the user to change any existing parameters for post-flashover burning. *Modify Vents* allows the user to change the size of any predefined vent. *Stop* terminates the simulation and returns to the fire scenario overview window.

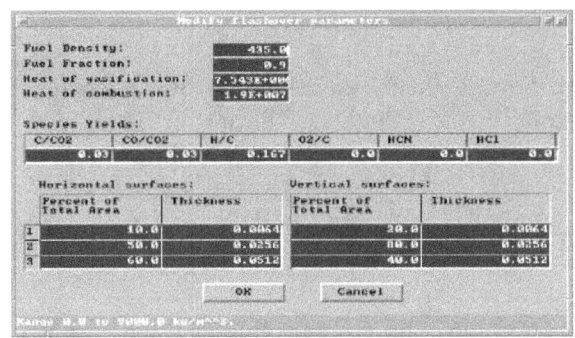

Flashover Clicking the *Modify Parameters* button displays the flashover screen. The flashover screen has a number for entries of information needed to continue the simulation in flashover. Default values are provided for every field.

Fuel density: Average density of the fuel burning during post-flashover burning

Fuel fraction: the fraction of the total material that will pyrolyze and lead to combustion.

Heat of gasification: the amount of energy necessary to pyrolyze a unit mass of the fuel – that is turn it from a solid to a gas.

Heat of combustion: the amount of energy a given mass of the fuel gives off when it burns.

Combustion chemistry: The next row of entries defines the combustion chemistry for the fuel. *C/CO2* field is the ratio of mass of carbon produced for smoke to the mass of carbon dioxide produced in combustion of the fuel. *CO/CO2* is the ratio of mass of CO produced to the mass of CO2 produced in combustion of the fuel. *H/C* is the ratio of the mass of hydrogen to carbon in the fuel. *O2/C* is the mass of oxygen liberated to the atmosphere by pyrolization to the mass of carbon in the fuel. *HCN* and *HCl* are the ratio of the mass of HCN and HCl produced to the mass of CO2 in combustion.

Available fuel: These tables are used to put in the amount of fuel available in the room. One table is for fuel with its surface oriented horizontally; the other is for fuel with its surface oriented vertically. For each orientation, up to three thicknesses of fuel can be specified. The amount of fuel is given as a percentage of the total room surface with the same orientation. For example, for horizontal fuel, this is the percentage of the total wall surface area. The area in a table can add up to more than 100% to allow for the inclusion of surface area of furniture, partitions, or other material in the compartment. When all the inputs are satisfactory clicking on the *OK* button continues the run.

3.4 Saving the Results of the Simulation

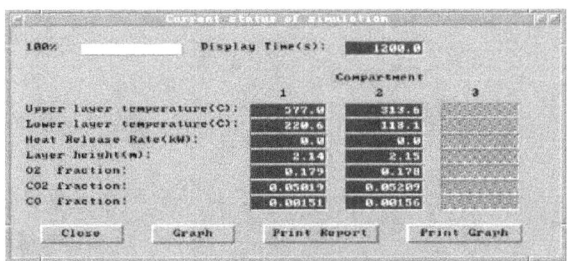

When the simulation is complete, the results of the calculations are available in several forms. If spreadsheet output was selected (see section 2.2.4 for details on specifying spreadsheet output), the user-specified file can be imported into a spreadsheet or other program for further analysis. In addition, the results can be printed in tabular or graphical form. Four buttons are available at the bottom of the simulation status window once the simulation is complete:

Close: Select *Close* to close the simulation status window (and associated graph window if displayed) and return to the fire scenario overview window. If *Close* is selected prior to the *Print Report* and/or *Print Graph*, below, the simulation must be repeated to be able to print the tabular report or graph.

Graph: The *Graph* button controls the display of a window showing graphs of the upper layer temperature, the calculated HRR, and the upper layer depth. Once the graph window is displayed clicking on the graph button closes the graph window.

Print Report: Select the *Print Report* button to produce a printed summary of the input file and the results of the simulation.

Print Graph: Select the *Print Graph* button to produce a printed copy of the three graphs displayed with the *Graph* button, above. Printed graphs are only available on printers which support the HPGL graphics language. This includes all Hewlett-Packard LaserJet printers beginning with the LaserJet III. For other printers, consult the printer documentation to see if your printer supports the HPGL graphics language.

4 An Advanced Example

This section provides an example using FASTLite which simulates an actual fire incident. Since the input is more complex than the example presented earlier, the user should be familiar with the use of FASTLite prior to working this example.

4.1 The Incident

On March 28, 1994, the New York City Fire Department (FDNY) responded to a report of smoke and sparks issuing from a chimney at a three story apartment building in Manhattan. The officer in charge ordered three-person hose teams to make entry into the first- and second-floor apartments while the truck company ventilated the stairway from the roof. When the door to the first-floor apartment was forced open, a large flame issued from the apartment and up the stairway, engulfing the three firefighters at the second floor landing. The flame persisted for at least 6½ minutes, resulting in their deaths. The FDNY requested the assistance of the National Institute of Standards and Technology (NIST) to model the incident in the hope of understanding the factors which produced a backdraft condition of such a duration.

The Building: The fire occurred in a three story, multiple brick dwelling of ordinary construction approximately 6.1 m (20 ft) wide by 14 m (46 ft) deep, and 3½ stories tall. The building contained four apartments, one on each story, with the basement apartment half below grade. While the basement apartment had its own entrance, access to the others was by an enclosed stairway running up the side of the building. The building was attached to an identical building that was not involved.

The buildings were built in the late 1800's and had undergone many alterations over the years. Recent renovations included replacement of the plaster/lathe with drywall on wood studs, lowering the ceilings to 2.5 m (8.25 ft), new windows and doors, heavy thermal insulation, sealing and calking to minimize air infiltration (the building was described as very tight.). Built before central heat, the apartments had numerous fireplaces, most of which had been sealed. The apartment of fire origin had two fireplaces, but only the one in the living room was operable. All apartments had thick plank wood floors.

The apartments had similar floor plans; the differences resulting from the stairway. A floor plan of the first floor apartment is presented in figure 2. There was a living room in the front, kitchen and bathroom in the center, and a bedroom in the rear. Not found in the other apartments, the first floor apartment had an office within the bedroom which was not significant in the fire. The roof had a scuttle for access and a wired glass skylight located over the stairway.

The Fire: On March 28, 1994 at 7:36 p.m., the New York City Fire Department received a telephone report of heavy smoke and sparks coming from a chimney at 62 Watts St., Manhattan. The initial response was 3 engines, 2 ladders, and a battalion chief. On arrival they saw the smoke from the chimney but no other signs of fire. The engine companies were assigned to ventilate the roof above the stairs by opening the scuttle and skylight, and two three-person hose teams advanced lines through the main entrance to the first- and second-floor apartment doors.

The first-floor hose team forced the apartment door and reported:

> a momentary rush of air into the apartment, followed by
> a warm (but not hot) exhaust, followed by
> a large flame issuing from the upper part of the door and extending up the stairway.

The first-floor team was able to duck down under the flame and retreat down the stairs, but the three men at the second-floor level were engulfed by the flame which now filled the stairway. An amateur video was being taken from across the street and became an important source of information when later reviewed by the fire department. This showed the flame filling the stairway and venting out the open scuttle and skylight, extending well above the roof of the building. Further, the video showed that the flame persisted at least 6½ minutes (the tape had several pauses of unknown duration, but there was 6½ minutes of tape showing the flame).

Damage to the apartment of origin was limited to the living room, kitchen and hall -- closed doors prevented fire spread to the bedroom, bath, office, and closets. There was no fire extension to the other apartments and no structural damage. The wired glass in the skylight was melted in long "icicles" and the wooden stairs were mostly consumed. The description provided by the surviving hose team was of a classic backdraft; but these usually persist only seconds before exhausting their fuel supply. Where did the fuel come from to feed this flame for so long?

Cause and Origin: The subsequent investigation revealed that the first-floor occupant went out at 6:25 p.m., leaving a plastic trash bag atop the (gas) kitchen range which he was sure was turned off. It is reasonable that the pilot light ignited the bag, which then involved several bottles of high alcohol content liquor on the counter, and spread the fire to the wood floor and other contents. The occupant confirmed that all doors and windows were closed, so that the only source of combustion air was the fireplace flue in the living room from which the smoke and sparks were seen to emerge.

4.2 Computer Analysis

Clearly, the fire burned for nearly an hour under severely vitiated conditions. The open flue initially provided expansion relief and later vented smoke as the ceiling layer dropped below the level of the opening. Such vitiated combustion results in the production of large quantities of unburned fuel and high CO/CO_2 ratios. As shown in studies of the backdraft phenomenon, when a door is opened under such conditions, warm air flowing out is replaced by ambient air which carries oxygen to the fuel. When this combustible mixture ignites, a large flame extends from the door. To determine whether enough fuel could collect within the apartment to feed the flame for the period of time observed, FASTLite was used to recreate the incident.

The apartment of origin was modeled as a single room with dimensions of 6.1 m by 14 m by 2.5 m. The stairway was modeled as a second room 1.2 m by 3 m by 9.1 m connected to the apartment by a closed door and having a vent at the roof of 0.84 m^2 area. The fireplace flue was modeled as a small leakage opening at the height of the top of the fireplace opening.

The initial fire was assumed to be a constant heat release rate (HRR) of 25 kW from actual data on burning trash bags. This fire then transitioned into a "medium growth rate t-squared" fire with a peak heat release rate of 1MW; however this was reached only momentarily due to limited oxygen. Such "medium t2" fires are characteristic of most common items of residential contents.

4.3 Entering the Data

To start FASTLite, change to the directory which contains the program and run it by changing to the directory where the software was installed and executing the program. The following commands are typical:

```
cd \fastlite
fastlite
```

from the DOS command prompt (pressing the Enter key after each command). After the FASTLite logo appears, the main desktop menu is displayed. Using the mouse, click on *File* and *New* to begin the definition of a new fire scenario.

Once *New* is selected, the structure selection window is displayed. From this window, the number of compartments and the arrangement of the compartments is defined. Once an initial selection has been made, the size of the compartments and any connections between compartments will be customized to fit the details of this specific scenario. For this example, choose a "2 Compartment" example by using the mouse to select the "2 compartment" icon. Note that the selection highlights the icon with a bold black border. This selection defines two 2.4 m wide by 3.6 m deep by 2.4 m high compartments with connecting doorways. These dimensions will be customized to fit the current building geometry below. For now, click on *OK* to accept the selection and close the window.

For this example, a customized medium T-squared growth rate fire will be specified. In a mixed collection of fuels selecting the medium curve is appropriate as long as there is no especially flammable item present. In a manner similar to the selection of the 2 compartment structure geometry above, select the custom growth rate fire by using the mouse to select the "Custom Growth" icon. Note that the selection highlights the icon with a bold black border. Click on *OK* to close the window. A window is displayed to define critical time points for the fire. For this example, enter 300 s and 3000 s respectively. Click on *OK* to accept these values and close the window. This fire description will be further customized to fit the current fire scenario below.

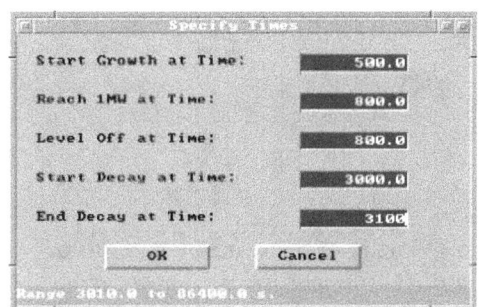

For this test case, typical metric units will be used. Select *Options* from the desktop menu, then selecting *User Specified Units* to display the measurement units window. To modify the settings for any of the base measurements, click on the pull-down icon to the right of the measurement. A pull-down menu is displayed, listing the available measurement units. Select the display units desired by clicking

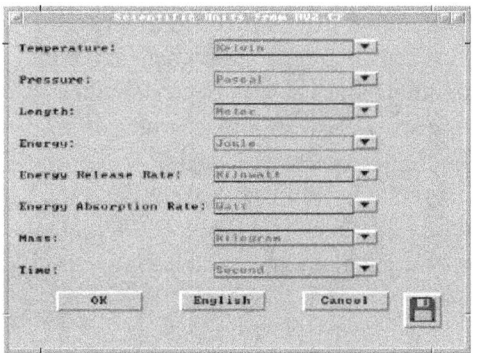
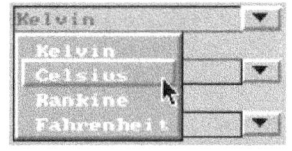

37

on the corresponding menu entry. For this test case, change the *Temperature* to Celsius and the *Energy Release Rate* to kilowatts.

4.3.1 Entering Title for the Input File

 Before beginning to define the structure, it is typically best to start by assigning a title or description to the input file. Tab to the title field or move the mouse cursor to this field and click with the left mouse button. Enter the text "62 Watts St Fire" as a title for this example.

4.3.2 Defining Ambient Conditions

 To edit the internal and external ambient temperature, pressure, and station elevation along with information on external wind, Tab or move the mouse to the ambient conditions icon in the environment section of the fire scenario overview window, and click or press Enter. For this example, change the internal ambient temperature to 20°C and the external ambient temperature to 10°C. Click on *OK* to accept these values and close the Ambient Conditions window.

4.3.3 Specifying Simulation Time and Spreadsheet Output

 To specify the simulation time, display time, or output to spreadsheet file with output interval, Tab or move the mouse to the filename icon in the environment section of the fire scenario overview window, and click or press Enter. For this example, enter 3000 s for the simulation time, 10 s for the display interval, 30 s for the spreadsheet interval, and WATTS.TXT for the name of the spreadsheet file.

 Save the file by selecting the diskette icon and select the *Save As* option. Enter the file name as watts.dat and click on *OK* to save the file.

4.3.4 Modifying Compartment Geometry

In order to model a fire scenario, the user must portray the geometry of the structure in terms of the size and elevation of every compartment in the structure. Thermophysical properties of the enclosing surfaces must also be specified by selecting surface materials in order to accurately model the transfer of heat through the surfaces. When the new data file was defined, default values for the compartment size, surface materials, and vent sizes were defined. These must be customized to fit each specific example. For this example, begin with the fire compartment by selecting the compartment and then the geometry icon to display the geometry window. Modify the compartment sizes as follows: the depth to 6.1 m, the width to 12.8 m, and the interior height to 2.53 m. The enclosing surfaces default to gypsum for the ceiling and walls and plywood for the floor. For this example, only the floor needs to be changed – selecting HARDWOOD from the pull-down list to the right of the floor edit widget. In a similar manner, modify the geometry for compartment (the stairway) as follows: the depth to 3.05 m, the width to 1.22 m, the height to 9.14 m, and the floor surface to HARDWOOD. Click on *OK* to accept these values and close the Geometry window.

Save the file by selecting the diskette icon and select the *Save* option.

4.3.5 Modifying the Vent Connections

 To specify a horizontal flow opening, select the corresponding compartment in the structure graphics list on the fire scenario overview window. When the new data file was defined, several vents were included – doors between the two compartments and the outside, and typical leakages to the outside. These must be customized to fit each specific example. Begin by selecting the fire

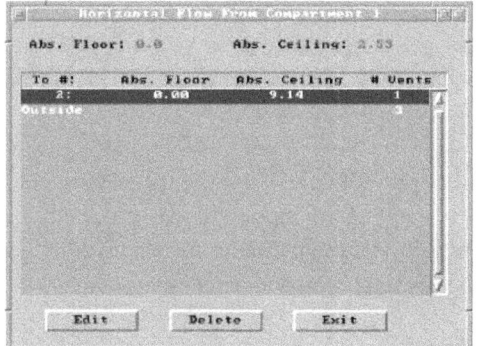 compartment and Tab or move the mouse to the horizontal flow icon, and click or press Enter. The vent connection window is displayed. Three default vents are shown. Begin by selecting the connections to compartment 2 and select Edit to change the vents to the stairway. For this example, simulate the closed apartment door by changing the soffit height to 0.000213 m – a tiny crack at the bottom of the doorway. This vent will be modified during the simulation to account for the doorway opened by the firefighters at 2248 s. Click on *OK* to accept these values and close the edit window.

In a similar manner, edit the vents to the outside. Two vents to the outside are displayed in the Horizontal Flow window. For this example, it will be assumed that the apartment was tightly sealed to eliminate air infiltration. The only vent to the outside from the apartment will be assumed to be the working chimney in the living room. Select Outside and the Edit to display the existing vents to the outside. Press *Alt-D* to delete the first predefined vent. The last vent will by customized to simulate the fireplace opening. For this opening, define the width to be 1.321 m, the sill to be 1.1 m, and the soffit to be 1.143 m. Click on *OK* to accept these values and close the edit window. Click on *OK* again to close the vent connection window and return to the fire scenario overview window.

In a similar manner, define two vents from the stairway (compartment 2) to the outside. The exterior door at the bottom of the stairway is 0.91 m wide by 2.13 m high will a sill height of 0.0 m. To simulate the roof vent, a vent is placed near the ceiling with a cross sectional area of 0.84 m^2. Define this vent to be 3.05 m wide with a soffit height of 9.10 m and a sill height of 8.869 m.

 Save the file by selecting the diskette icon and select the *Save* option.

4.3.6 Modifying the Fire Definition

 A medium "t-squared" fire appropriate for this example was defined earlier. To finish the definition, the fire must be customized to position the fire and add the burning trash bags as the first item ignited. The fire position will be modified first. Select the fire icon to display the main fire specification window. For this example, change the fire position to an X position of 2.15 m and a Y position of 1.65 m to approximate the position of the kitchen stove in the apartment. Click on *OK* to accept these values and close the main fire specification window.

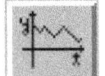 To modify the heat release rate curve for the fire to include the burning trash bags, select the fire compartment from the fire scenario overview window and the time curve icon. Select heat release rate from the menu to display the heat release rate window. Customize the heat 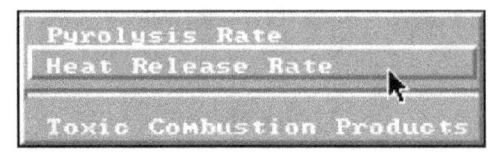 release rate curve for this example by including the heat release rate of the burning trash bags as follows: Change the heat release rate for the first two time points (t=0 s and t=500 s) to 25 kW. Click on *OK* to accept these values and return to the fire scenario overview window.

This completes the definition of the test case.

4.3.7 Saving Data File Modifications

Before the simulation can be run, the user must request the input editor to store the information on the computer's hard disk. All information for the current session of the input editor is grouped together on the hard disk and assigned a unique name referred to as the filename for retrieval purposes in the future. For this example, tab or move the mouse to the diskette icon and select the *Save* option.

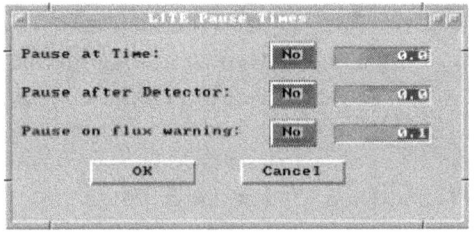

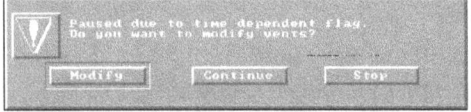

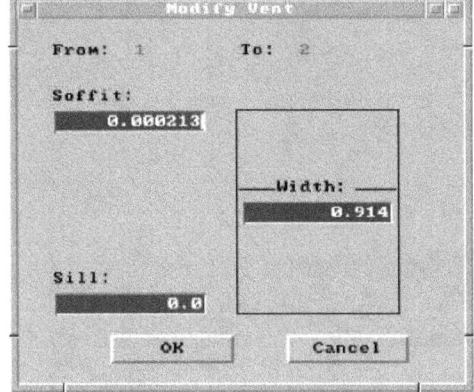

4.3.8 Running the Simulation

 To run the simulation, use the mouse to select the *Run Simulation* icon.

Before beginning the simulation, a window is displayed to provide the user with the option to stop the simulation at a time relative to the beginning of the simulation or to the operation of a detection device. For this example, enter a pause time of 2250 s so that the doorway between the apartment and stairway can be opened. Click on *OK* to accept these inputs and begin the simulation.

When the simulation is paused 2250 s, select *Modify* to display a window which allow the characteristics of the vents to be modified. Select the first entry in the list – the now closed doorway from the apartment to the stairway. Open the doorway by changing the soffit height to 2.13 m. Select *OK* then *Continue* to continue running the simulation.

4.4 Results of the Simulation

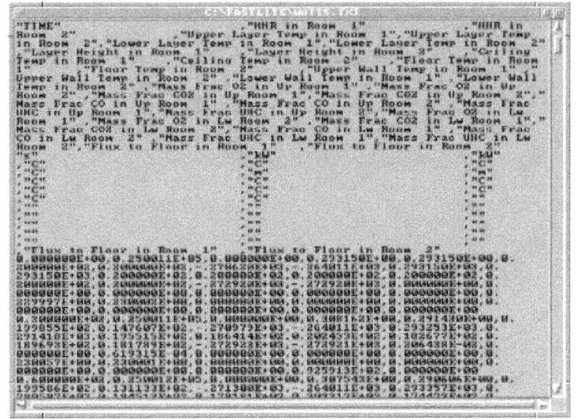

The spreadsheet output of the simulation, shown in the window to the left, was used to produce the graph shown below. From the spreadsheet data, an analysis of the simulation can be made.

The fire grew briefly to nearly 1 MW over 5 minutes of simulation time, then rapidly throttled back as the oxygen concentration dropped below 10%. Temperatures in the apartment peaked briefly at about 300 °C at the time of peak burning, then rapidly dropped below 100 °C as the burning rate fell. The concentration of carbon monoxide (CO) rose to about 3000 ppm and unburned fuel accumulated within the apartment volume during this stage of vitiated combustion.

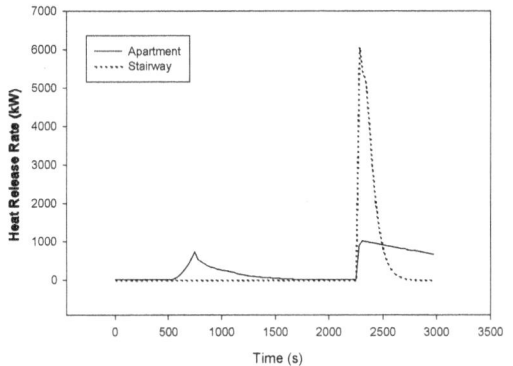

The front door was opened at about 2250 seconds into the simulation as an estimate of when the first floor team made entry. Immediately, there was an outflow of warm (100 °C) air from the upper part of the doorway, followed by an inrush of ambient air in the lower part of the doorway, followed by the emergence of a large door flame -- exactly as reported by the firefighters. This door flame grew within a few seconds to a peak burning rate of more than 5 MW, raising the temperature in the stairway to over 1200 °C – sufficient to melt the glass in the skylight, as observed. Most importantly, the quantity of unburned fuel accumulated in the apartment caused the door flame to persist for more than 7 minutes.

The FASTLite calculations showed that the theory of the development of this fire was technically sound. The calculations supported the hypothesis that unburned fuel and CO accumulated in an apartment with an open fireplace flue but otherwise tightly sealed, resulting in a backdraft on opening of the apartment door. They showed that sufficient fuel could accumulate under these under ventilated conditions to cause the door flame to persist for the extended period observed. Reported conditions such as flows observed in the doorway, melting of the glass skylight, and fire damage in the apartment and stairway, were consistent with the model results.

5 Estimation Tools

 Quick estimation tools derived from FIREFORM are available in the FASTLite GUI shell using one of two selection options. Select the *Tools* entry from the desktop menu, or select a compartment on the fire scenario overview window, and click on the tools icon. An input specification window for the selected calculation is displayed. If the calculation tool was selected using the tools icon on the fire scenario overview window, default values are obtained from current input specifications. If the calculation tool was selected from the desktop menu, no default values are available, and the user is required to supply all details. Procedures are discussed in detail in the following subsections. Users unfamiliar with the concepts of a Graphical User Interface (GUI) are strongly encouraged to review these concepts in the Getting Started section of this guide before continuing with the Estimation Tool section. Familiarity with GUI terminology and the use of a mouse is assumed throughout the remainder of this section.

For comparison purposes, the original FIREFORM which accompanied FPETOOL version 3.2 is available from the desktop menu.

The following discussion assumes strict SI units. However, actual data input is dependent upon the measurement units selected by the user.

5.1 Egress Time

This procedure estimates the time needed for a person or group of people to exit an area. The egress movement may be vertical or horizontal and include the use of doorways, stairs, ramps, and corridors. Elevator transportation is not considered.

Theory The egress time calculation assumes that evacuees will travel at user-designated speeds on flat and vertical pathways. These speeds will be altered if the user designates a reduced travel efficiency for the slowest person in the evacuating population. The default travel speed on flat pathways is 76.2 m/min (250 ft/min). The default travel speed on stairs is 12.2 m/min (40 ft/min). Travel time on vertical pathways may be further altered by deviations in standard-stair design measurements. The assumed standard-stair design measurement is a tread depth of 280 mm (11 in.) and a riser height of 178 mm (7 in.). The base speed on stairs is increased for several user-defined parameters. These parameters are increased tread depth, decreased riser height and increased effective width [16] of the stairway. The assumed exitway flow rate is 60 persons/min/$m_{W\text{-effective}}$ (18.3 persons/min/$ft_{W\text{-effective}}$). The rate of travel through enclosed exitways is limited by the flowrate through the doors or door-leaves in the enclosure opening. The doorway calculations assume a default movement rate of one person per second per door-leaf. The standard door-leaf width is 0.76 m (30 in.).

In the equations below, eq (1) represents the time needed for one individual to complete unimpeded egress. Equations (2) - (4) support eq (1). Equation (5) represents the time to move the entire building population through the exterior exit doors. Equation (6) represents the time to move the entire building population through and out of the stairway enclosures. In eq (6) the limit to flow is the $W_{effective}$. The $W_{effective}$ may be either the stairway enclosure exit door width or it could be the width of the stairway itself (protruding handrails or other projections).

$$t_{unimpeded} = \frac{(t_{horizontal} + t_{vertical})}{\chi_{mobility}} \quad (1)$$

$$\chi_{mobility} = \frac{X}{100} \quad (2)$$

Together, eqs (1), (5) and (6) provide a first-order estimate of area evacuation times; the user, however, should be aware of assumptions in arriving at the results. The egress estimates assume the most efficient exit paths are chosen. The procedure does not account for investigation, verification, "way-finding," or assistance. Flow is assumed to proceed ideally and without congestion. There are no adjustments to flow speed in response to evacuee flow density. In light of these inefficiencies, it would be reasonable to expect evacuation times to be two to three times greater than the nominal evacuation time [17]. The nominal evacuation time varies. If any evacuation time from eqs (1), (5) or (6) was an order-of-magnitude greater than the other two evacuation times, then this would be the nominal evacuation time estimate. If the unimpeded evacuation time, ($t_{unimpeded}$) is close to one or both estimates of eqs (5) or (6), then $t_{unimpeded}$ plus a fraction of eqs (5) or (6) is the nominal evacuation time. Conversely, if eqs (5) or (6) exceeds $t_{unimpeded}$, then the nominal evacuation time is the time in eqs (5) or (6) plus a fraction of $t_{unimpeded}$. To determine what value these fractions should be, it is necessary to conduct a more detailed analysis of the evacuation flow [17].

$$t_{horizontal} = \frac{x_{horizontal}}{v_{able}} \quad (3)$$

$$t_{vertical} = \frac{z_{vertical}}{v_{stair}} \sqrt{\frac{11}{7} \frac{z_{riser}}{x_{tread}}} \quad (4)$$

$$t_{exit-opening} = \frac{N_{people}}{N_{exitleaves}} \left(\frac{exit\,leaf \cdot sec}{1\,person}\right) \quad (5)$$

$$t_{stair} = \frac{N_{people}}{W_{effective}} \frac{1}{\dot{Q}_{stair}} \quad (6)$$

$N_{exit\,leaves}$	Total number of door leaves from the building to the outside
N_{people}	Total evacuating population
$\dot{Q}_{stair}$	People flow rate in a stairway enclosure (default 60 people/min/m $_{W,\,eff}$)
t	Exit time (sec)
$W_{effective}$	Effective width of an exit passageway (see Section 3.6.3) (m)
$x_{horizontal}$	Total horizontal distance traversed by the evacuee (m)
v_{able}	Speed of an able evacuee moving on flat, dry surface (m/s)
v_{stair}	Speed of an able evacuee moving in a vertical means of egress (m/s)
x_{tread}	Depth of the tread from riser to riser (m)

X	Speed of the slowest evacuee as a percentage of able evacuee speed
z_{riser}	Height of the riser from tread to tread (m)
$z_{vertical}$	Total vertical traverse distance (not distance along a sloped incline) (m)

Notes Door leaves less than 5/6 of a standard-door width (0.76 m) should not be considered as an additional leaf available for evacuee egress movement.

Flow rates through door leaves are assumed at one person per second per door leaf. If the door leaf is less than 0.86m (34 in.) then the flow rate may be less. The exitway flow rate is user adjustable. For exit openings substantially larger than 0.86m (34 in.) per door leaf, the flow rate can exceed one person per second. To reflect this potential the user should modify the parameter "Flow rate per door leaf."

Effective flow width for stairs measures wall-to-wall, minus projection of artifacts, minus a clearance distance from the artifacts. This clearance distance is artifact dependent [3]. Typical stairwell effective width is 0.305 m (1 ft) less than actual width. This accounts for the 0.076 m (3 in.) projection of each handrail plus 0.076 m clearance for each handrail.

Turnstiles can have flow rates 1/3 of values for stairwell doorways (20 people/min/$m_{W,\ effective}$).

Stair flow rate is roughly 60 persons per minute per meter of effective width (18.3 persons/min/ft). This flow rate is user adjustable.

If there is more than one stairway and these widths differ, an average width is needed to represent these egress paths because only one stairway width may be entered to this routine. This average width may be calculated such that when multiplied by the number of stairways, it yields a net width equal to the sum of the individual stairway widths.

Emergency travel speed on flat, dry, uncongested surfaces is 76 m/min (250 ft/min). The flow rate is user adjustable; one application follows:

The Americans with Disabilities Act [2] suggest flow rates of 28 m/min (90 ft/min) for disabled evacuees. This represents a speed 37 percent of that assumed for an able person. However, after every 30.5 m (100 ft) of travel, the ADA further suggests that the evacuee will pause for 2 minutes, presumably to rest.

The output parameter "Time required to pass persons through the (building) exit doors..." predicts the waiting time the **last** person in line will experience when completing one of the last parts of their egress path--that of moving from inside to outside of the building. The rate of movement through the exterior-access doors is a function of the number and effective width of these doors. The building population is equally distributed among the specified exterior-access exit doors. A wait during this part of the egress can be caused by insufficient exit doors, inadequate door widths, or a larger-than-anticipated building population.

The output parameter "Time required to pass persons through the stairwell exit doors..." is calculated analogous to "Time required to pass persons through the (building) exit doors..."

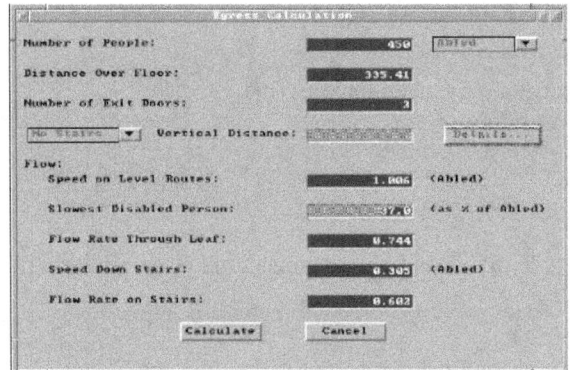

Execution Select *Tools* from the desktop menu or click on the tools icon. Select *Egress Time* from the tools menu. The egress time window is displayed.

Note the display of range and measurement units at the bottom of the window as each edit widget is made the focus widget. To customize the units see the Advanced Features section 6.1.

Number of People: Total number of evacuees using the evacuation routes. Evacuees may be considered "abled" or disabled which modifies the movement speed by the "Slowest Disabled Person" input below.

Distance Over Floor: Total travel distance over level surfaces.

Number of Exit Doors: Total number of exit doors leaves available to evacuees. This number is an integer. To obtain results reflecting additional exit width beyond 0.86m (34 in.) adjust the parameter pertaining to "Flow Rate Through Leaf."

Vertical Distance: Vertical distance moved via stairwell travel. This is not the same as the distance moved along the slope of a stairway but the vertical distance between the starting and the stopping locations. Note that this input is only available if stairs are selected by the pulldown widget. Additional details of the stairway calculation can be input by selecting *Details...* The number of separate stairways, width of stairways (average or total width), riser height, and tread depth may be input.

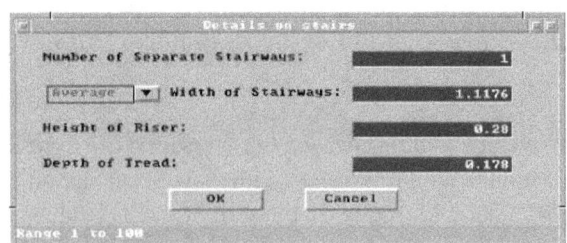

Speed on Level Routes: Travel speed of able evacuees over level surfaces.

Slowest Disabled Person: Travel speed of the slowest disabled evacuee expressed as a percentage of the travel speed of able evacuees.

Flow Rate Through Leaf: Travel speed of evacuees through exit doorways, expressed as persons per unit time per exit door leaf.

Speed Down Stairs: Travel speed of able evacuees on stairways.

Flow Rate on Stairs: Flow rate of evacuees on stairways expressed as persons per unit time per unit width of stairway.

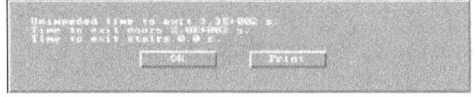

Press the *Calculate* button to begin calculation. Upon completion, a results windows is displayed. The "horizontal and stair travel time" is the time estimated for a person to traverse all stair and horizontal paths exclusive of any queuing. Doorways are assumed open and no other evacuees are considered to impede travel rates. The "time required to pass persons through the (building) exit doors" is the time for the entire building population to pass through the available building exit doors. The "time required to pass persons through the stairway exit doors" is the time for the entire building population to pass through the available building stairwell exit doors.

5.2 Sprinkler / Detector Activation

This procedure calculates the thermal response of a detector or sprinkler located at or near a ceiling whose area is large enough to neglect the effects of smoke layer development.

Theory The equations in this procedure were originally distributed in a program written by Evans and Stroup [18] entitled DETACT-QS. The correlations for jet temperatures and velocities were developed from data by Alpert [19]. The theory and documentation for sprinkler activation are presented by Evans [20].

$$T_{D,t+\Delta t} = T_{D,t} + (T_{jet_{t+\Delta t}} - T_{D,t})(1 - e^{-\frac{1}{\tau}}) + (T_{jet_{t+\Delta t}} - T_{jet_t})\tau(e^{-\frac{1}{\tau}} + \frac{1}{\tau} - 1) \tag{1}$$

The results of this procedure predict time of thermal detector activation. In order to make this prediction, time-dependent events from the fire must be linked to events resulting in the heating of the detector from ambient to its activation temperature. The heat source is accounted for by a user-specified, time varying-fire. The time-lag associated with heating the detector is accounted for with the RTI parameter, eq (2). The RTI parameter considers the detector's ability to absorb heat and the ambient environment's ability to provide heating. Ambient environmental heating is modeled with only forced convection. The temperature and velocity of the convecting air, eqs (3) - (4), are correlations assembled from experimental data of full-scale steady-state and growing fires [19]. The actual heat release rate of the fire should be used with both radiative and convective fractions. When $T_{D,t+\Delta t}$ equals or exceeds the value in $T_{D, activation}$, then detector response is predicted.

$$\tau = \frac{RTI}{\sqrt{v_{jet_t}}} \tag{2}$$

$$v_{jet_t} = 0.95\left(\frac{\dot{q}}{z}\right)^{\frac{1}{3}}, \quad for \; \frac{r}{z} \leq 0.15$$

$$= 0.2 \frac{\dot{q}^{\frac{1}{3}} z^{\frac{1}{2}}}{r^{\frac{5}{6}}}, \quad for \; \frac{r}{z} > 0.15 \tag{3}$$

$$T_{jet_t} = T_\infty + \frac{16.9 \dot{q}^{\frac{2}{3}}}{z^{\frac{5}{3}}}, \quad \text{for} \quad \frac{r}{z} \leq 0.18$$

$$= T_\infty + \frac{5.38}{z}\left(\frac{\dot{q}}{r}\right)^{\frac{2}{3}}, \quad \text{for} \quad \frac{r}{z} > 0.18 \tag{5}$$

$\dot{q}$	Total theoretical fire heat release rate
r	Radial distance of the sprinkler from the vertical axis of the fire
RTI	Response Time Index: $(mc_p)_{detector}/(hA) \cdot v^{\frac{1}{2}}_{jet}$, a characterization of the detector's thermal sensitivity; a measure of how quickly a detector link reaches its activation temperature.
$T_{jet,t+\Delta t}$	Temperature of the jet at the next time step, t+Δt
$T_{jet,t}$	Same as $T_{jet,t+\Delta t}$, but at the previous time step, t
T_∞	Ambient space and initial sprinkler temperature
$T_{D,t}$	Detector or link temperature at time, t
$v_{jet,t}$	Velocity of the ceiling jet gases as a function of the parameters on the right-hand side of eqs (2) and (3) at time step, t
z	Vertical entrainment distance; the difference between the height of the ceiling and the base of the flames

Notes The total theoretical heat release rate should describe the fire in this correlation [19], [21]. The total theoretical heat release rate may be obtained by multiplying the mass pyrolysis rate by the theoretical heat of combustion. Mass pyrolysis rates can be obtained through experimental measurement using load cells or by analogy with previously burned exemplars. Theoretical heats of combustion are available from handbooks.

The program assumes a quasi-steady-state fire and ceiling jet behavior. This assumption limits the accuracy most when the fire heat release rate changes very rapidly.

The procedure assumes an unconfined ceiling jet and plume. If a smoke layer should develop under the ceiling (as is the case when the fire is large relative to the room), the fire model within FASTLite will consider entrainment of hot gases into the fire plume whereas this procedure will not.

If the detector is located significantly below the bottom of the ceiling jet, then this procedure should not be used. The ceiling jet thickness is estimated between 6 - 12% of the entrainment height (z).

The procedure assumes the detector is located such that it is exposed to both the maximum ceiling jet velocity and temperature.

Correlations for ceiling jet temperature and velocity were determined from limited experimental data: (no beam or truss ceilings, no cathedral ceilings, only smooth, horizontal unconfined ceilings).

Sprinklers or heat detectors located on a wall or on a ceiling next to a wall may have activation times significantly later than predicted activation times. This delay is due to the dissipation of ceiling jet velocity at the wall and wall/ceiling intersection. The dissipation effect may be especially true in room corners.

Rate-of-rise heat detectors are not simulated, only fixed-temperature detectors are simulated.

Radiation and conduction are not accounted for explicitly, but because these phenomena participated in the correlational experiments--to the degree that simulated fire conditions reproduce experimental fire conditions--the radiation/conduction effects are implicitly accounted for. The experimental fire conditions involved wet-pipe sprinklers exposed to cotton, wood, polyurethane, polyvinyl chloride, and liquid heptane fires [18], [22].

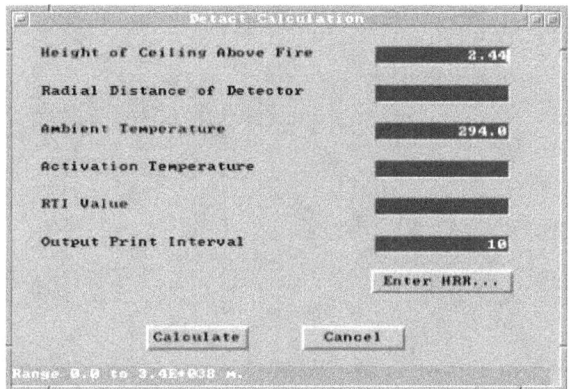

Execution Select *Tools* from the desktop menu or click on the tools icon. Select *Detector/Sprinkler Activation* from the tools menu. The detact calculation window is displayed.

Note the display of range and measurement units at the bottom of the window as each edit widget is made the focus widget. To customize the units see the Advanced Features section 6.1.

Height of Ceiling Above Fire: The difference between the elevation of the lowest point of the fire that can freely entrain air and the elevation of the ceiling.

Radial Distance of Detector: Horizontal distance of the detector location from a vertical axis running through the center of the fire

Ambient Temperature: T_∞, Ambient and initial sprinkler temperature

Activation Temperature: $T_{D, activation}$, Detector link activation temperature

RTI Value: Detector response time index. For sprinklers, this is now often given in the manufacturer's catalog. For fixed-temperature heat detectors, The table 1 provides values for devices which correspond to UL spacings. For rate-of-rise detectors, table 2 gives RTI values for three rate-of rise settings (note that 15 °F is typical for U.S. commercial devices).

Output Print Interval: Time interval between hard copy output of time, heat release rate, ceiling jet temperature, and sprinkler / detector link temperature.

Table 1. RTI values for fixed temperature heat detectors. RTI values shown are in $(ft-s)^{½} / (m-s)^{½}$. Since the original work was in English units, SI units are approximate.

UL Listed Spacing	UL Listed Activation Temperature						All FM Listed Temps.
	128°F	135°F	145°F	160°F	170°F	196°F	
10/3.1	894/494	738/408	586/324	436/241	358/198	217/120	436/241
15/4.6	559/309	425/235	349/193	246/136	199/110	101/56	246/136
20/6.1	369/204	302/167	235/130	158/87	116/64	38/21	157/87
25/7.1	277/153	224/124	174/75	107/59	72/40	---	107/59
30/9.2	212/117	179/99	136/75	81/45	49/27	---	81/45
40/12.2	159/88	128/71	92/51	40/22	---	---	---
50/15.3	132/73	98/54	67/37	---	---	---	---
70/21.4	81/45	54/30	20/11	---	---	---	---

Note: These RTI's are based on an analysis of the Underwriters Laboratories and Factory Mutual listing test procedures. Plunge test results on the detector to be used will give a more accurate response time index.

Table 2. RTI values for rate-of-rise heat detectors. RTI values shown are in $(ft-s)^{½} / (m-s)^{½}$. Since the original work was in English units, SI units are approximate.

UL Listed Spacing (ft/m)	UL Listed Activation Rate of Temperature Rise		
	15°F/min / 8°C/min	20°F/min / 11°C/min	25°F/min / 14°C/min
10/3.1	1834/1013	1308/722	984/543
12.5/3.8	1453/802	1073/593	805/445
15/4.6	1185/654	872/482	637/352
20/6.1	872/482	581/321	425/235
30/9.2	559/309	380/210	280/155
40/12.2	447/247	291/161	206/114
50/15.3	425/235	246/136	161/89

Enter HRR...: The heat release rate of the fire as a function of time is entered in a manner similar to the fire for the fire model. Using the mouse, click on the *Enter HRR...* button. The fire curve window is displayed. Note the display of range and measurement units at the bottom of the window as each cell in the edit list widget is made the focus widget. To customize the units see the Advanced Features section 6.1.

If the calculation was selected from the fire scenario window, heat release rate information for the fire is displayed. If no other time-dependent curves have been defined previously for this input file, no entries will be displayed in the spreadsheet. To enter heat release rate time curves for the first time, Tab to the spreadsheet or click on the first cell using the mouse. Enter the actual time at which each fire property is evaluated. Use the vertical scroll bar to enter additional entries beyond those initially displayed. Refer to the GUI Terminology section 1.2.2 of this reference guide for an explanation of the use of scroll bars in the GUI interface.

Movement between time cells is handled using the ↑ and ↓ keys. To insert a new time entry between two existing rows, press Alt-I. To delete a row, press Alt-d. Deleting a row moves all rows following the deleted row up one row. If the entry is to be erased, but all other rows are to remain in current positions, press Alt-e to erase the row. Time entry value ranges are constrained by the entries in surrounding time cells so that time entries within a column of the spreadsheet are ever increasing values. The maximum value allowed is 86400 seconds.

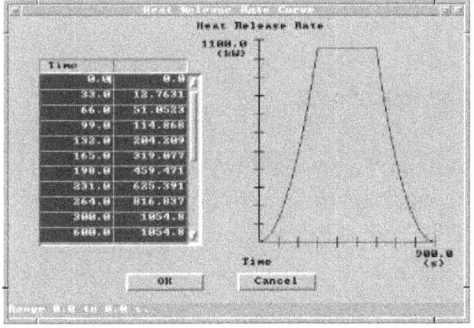

Once the time points have been specified in the first column of the curve spreadsheet, select the top cell of the right column to begin entering the heat release rate curve. Keyboard movement from the time column can be handled using the Shift-→ and pressing the ↑ key until the first cell is reached. Specify the values for the combustion property according to details discussed below. As values are entered in the right column, the fire curve corresponding to time and y-axis entries is plotted on the graph to the right of the spreadsheet.

5.3 Atrium Smoke Temperature

This procedure estimates the average temperature in the smoke layer developing from a fire within an atrium or other large space.

Theory The atrium smoke temperature is derived from the **ASETBX** plume equation [23] that had its own origins from Zukoski [24]. Given an entrainment height and a fire heat release rate, the procedure determines the maximum temperature in the plume.

$$T_{atria} = \frac{220}{1 + 39.8 \left(\frac{z^{5/3}}{\dot{q}^{2/3}} \right)} \tag{1}$$

T_{atria} — Temperature rise in atria hot gas layer (°C)
$\dot{q}$ — Fire heat release rate (kW)
z — Elevation between lowest point of entrainment and height of interest (m)

Notes The plume theory used in this routine does not apply to the case where plume gases expand to the point of contact with the compartment walls [25]. To ensure that the plume does not touch the wall in the modeled case, the following restriction may be reviewed. The restriction assumes a plume expansion angle of 15° from the vertical.

$$H_{room} \leq \frac{W_{room}}{2 \cdot \sin 15°}$$

Wall heat losses should be negligible. To ensure modeled wall heat-losses are negligible, either the plume should not touch the walls, or the smoke layer should have a temperature below 105 °C (220 °F). Cooper [25] suggests an upper limit on the fire heat release rate (kW) for maintaining moderate wall heat loss. The variable z is defined in eq (1). At fire sizes larger than Q_{limit}, heat from the gases lost to the walls can produce temperatures cooler than predicted.

$$\dot{Q}_{limit} = 333 \, z^{5/2}$$

The heat release rate is steady-state.

The fire is assumed to be a point-source; i.e. no line fires.

The program does not accurately model small fires and/or short entrainment heights.

Execution Select the compartment to be modeled from the fire scenario overview window, or select *Tools* from the desktop menu. If a compartment is selected from the fire scenario overview window, click on the tools icon. Select *Atrium Smoke Temperature* from the tools menu. The following window is displayed:

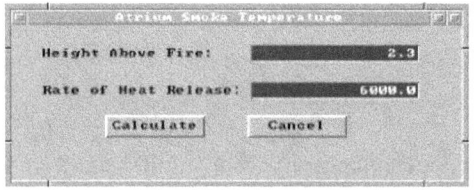

Note the display of range and measurement units at the bottom of the window as each edit widget is made the focus widget. To customize the units see the Advanced Features section 6.1.

Height Above Fire: Entrainment distance from the point where entrainment begins, usually the base of the flames, to the elevation of interest. If this procedure is selected for a compartment on the fire scenario overview window, a height corresponding to the ceiling of the selected compartment is provided as a default value.

Rate of Heat Release: Fire heat release rate.

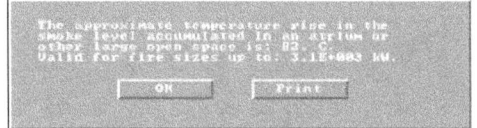

Press the *Calculate* button to begin calculation. Upon completion, a results window is displayed. The approximate plume temperature rise at the user specified height is reported. In addition, the largest actual fire heat release rate that can be used with this correlation for valid temperature approximations is provided.

5.4 Buoyant Gas Head

This procedure calculates the pressure difference between two laterally adjacent gases of different density. In fire safety applications, these density differences are created by differences in smoke and clean air temperatures, but the density differences could also be due to differences in molecular weights of adjacent gases.

Theory The equation used is directly extracted from the manual *Design of Smoke Control Systems in Buildings* [26]. The ambient, colder gas volume is assumed to be 21 °C (70 °F). The pressure differential is

$$\Delta P = 3460 \left(\frac{1}{294} - \frac{1}{T_h} \right) z \qquad (1)$$

calculated between the adjacent gases at an elevation coincident with the base of the least dense gas volume.

ΔP Pressure difference between the cold and warm gas (Pa)
T_h Temperature of the hotter gas (K)
z Thickness of the least dense (hot) gas volume (m)

Notes Conditions are steady-state. Temperature in the hotter, least dense, gas layer is uniform throughout, and the height of the hotter gas volume is constant.

There are no mechanical ventilation/pressurization connections with the hot layer.

Air is the surrounding fluid with a density of 1.2 kg/m³ (0.075 lb/ft³) at 21 °C. Use of this formula in environments where the surrounding gas has a molecular weight or pressure substantially different than a warm-gas layer of air at standard pressure will result in errors.

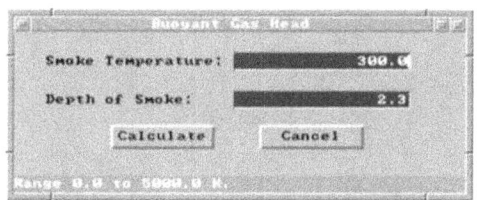

Execution Select the compartment to be modeled from the fire scenario overview window, or select *Tools* from the desktop menu. If a compartment is selected from the fire scenario overview window, click on the tools icon. Select *Buoyant Gas Head* from the tools menu.

Note the display of range and measurement units at the bottom of the window as each edit widget is made the focus widget. To customize the units see the Advanced Features section 6.1.

Smoke Temperature: Temperature of the hot, least dense, layer of gas. If this procedure is selected for a compartment on the fire scenario overview window, the previously specified internal ambient temperature is provided as a default value.

Depth of Smoke: Thickness of the warm air layer.

Press the *Calculate* button to begin calculation. Upon completion, a results window is displayed.

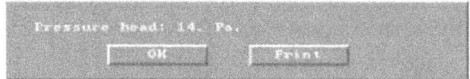

Pressure difference at the base of the hotter gas volume is reported.

5.5 Ceiling Jet Temperature

With this procedure's estimate of ceiling jet temperature one can determine the likelihood of ignition or heat-induced damage at locations *outside* the plume impingement zone.

Theory The likelihood of igniting ceiling combustibles may be determined from ceiling jet temperature estimates provided by this routine [27]. Gas jet temperatures are estimated for radial locations outside the plume impingement zone on the ceiling. The plume impingement radius is 0.2 of the plume clear entrainment height. The procedure will adjust the temperature of the gas to recognize the changed entrainment characteristics of wall- or corner-positioned fires. The entrainment adjustment is per the method of reflection [28] (4). The procedure also recognizes that any hot gas layer development beneath the ceiling will create an underestimate bias in the temperature predictions. As a precaution to this bias, the procedure approximates the time when this hot layer development will become influential [29].

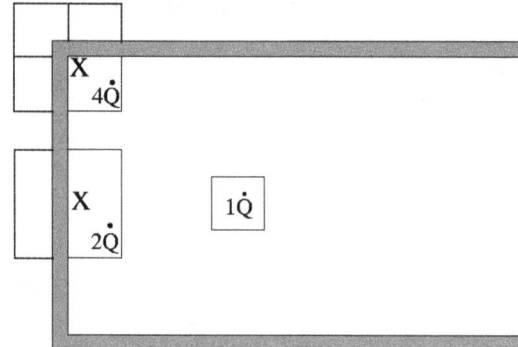

Method of reflection for fires in corners & against walls

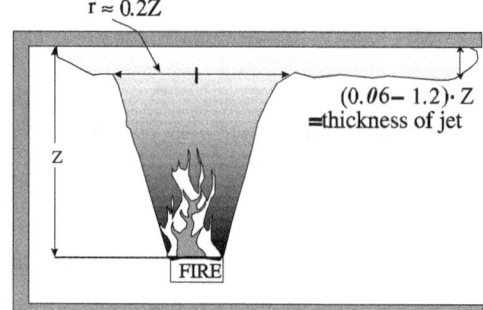

Characteristics of fire plume & ceiling jet

Fire plume and jet characteristics.

$$T_\infty + 6.81 \frac{(\frac{K\dot{Q}}{r})^{\frac{2}{3}}}{z} \; ; \quad for \; \frac{r}{z} > \quad (1)$$

$$t = \frac{20.3 \, A_{ceiling}}{\dot{Q}^{\frac{1}{3}} z^{\frac{2}{3}}} \quad (2)$$

A	Area (m²)
D	Fire diameter (m)
K	Entrainment factor (1 for axisymmetric, 2 for wall-fire, 4 for corner fire)
Q	Total theoretical fire heat release rate (kW)
r	Radial distance from center of the fire to point of interest outside the plume (m)
z	Vertical distance between ceiling and lowest point of the burning fuel (m)

t	Time (second)
T	Jet temperature at height, z, and radial distance, r, from the fire (°C)
T_∞	Ambient temperature (°C)

Notes The total theoretical, not the actual, heat release rate should be used to describe the fire in this and other ceiling/plume correlations by Alpert. The total theoretical heat release rate may be obtained by multiplying the mass pyrolysis rate by the theoretical heat of combustion. Mass pyrolysis rates can be obtained through experimental measurement using load cells. Theoretical heats of combustion are available from handbooks. This fire specification is documented per Alpert's work.

Points considered for examination should be at radial distances greater than 0.2 times the entrainment height from the vertical axis of the fire.

The entrainment height is the vertical distance between the ceiling and the lowest elevation where flaming combustion occurs.

The fire heat release rate is assumed to be steady state. The test fires used in developing the correlation were buoyancy dominated diffusion flames arising from various fuels: wood cribs and pool fires of liquid heptane and ethanol. This procedure is not intended for momentum-dominated jet fires.

The fire is assumed to be a point source; line fires are not considered.

The plume is considered to be unconfined up to the time of predicted layer development.

The method of reflection is appropriately used when flames are attached to a wall or a corner. When the fire is next to but not against the walls, and flames are not touching the wall surfaces, then the reduction in entrainment was not significant [3].

This routine does not consider combustible wall surfaces.

Standard pressure (101,325 Pa) and normal atmospheric gas concentrations (79% N_2, 21% O_2) exist.

This procedure is valid up to the point of hot gas layer development. The actual time to hot layer development can be less than predicted by eq (1) for values of $z/D_{fire} \gg 1$. In addition, eq (2) does not consider fires against a wall or in a corner. The estimated time to hot layer development is inappropriate in such circumstances.

Alpert developed two correlational predictions for ceiling jet temperatures. One correlation was to be used for predicting detector activation and the other for predicting thermal damage. The temperature predictions intended for detector-activation simulations are lower than the temperature predictions intended for thermal damage simulations.

Execution Select the compartment to be modeled from the fire scenario overview window, or select *Tools* from the desktop menu. If a compartment is selected from the fire scenario overview window, click on the tools icon. Select *Ceiling Jet Temperature* from the tools menu.

Note the display of range and measurement units at the bottom of the window as each edit widget is made the focus widget. To customize the units see the Advanced Features section 6.1.

Position of the Fuel: Select either *No nearby walls*, *Fuel package near a wall*, or *Fuel package in a corner*. Experiments have shown [30] that in order for reflection to apply, the fire flames must **touch** the wall.

Ambient Temperature: Compartment temperature at pre-fire conditions. If this procedure is selected for a compartment on the fire scenario overview window, the internal ambient temperature is provided as a default value.

Rate of Heat Release: Total theoretical fire heat release rate.

Distance From Fuel to Ceiling: Elevation difference between the lowest height where flames exist and where air can freely be entrained into the fire, and the height of the ceiling. If this procedure is selected for a compartment on the fire scenario overview window, a height corresponding to the ceiling of the selected compartment is provided as a default value.

Area of Ceiling: If this procedure is selected for a compartment on the fire scenario overview window, an area corresponding to the product of the compartment depth and width is provided as a default value.

Radial Distance From Fire Axis: Lateral distance across ceiling from a point directly over the fire to the point of interest in the ceiling jet.

Press the *Calculate* button to begin calculation. Upon completion, a results window is displayed.

The temperature in the gas jet at specified radial distance and height, and the time when a layer of hot gas that develops under the ceiling can interfere with the "unconfined ceiling jet" assumption are reported.

5.6 Ceiling Plume Temperature

This procedure estimates fire-plume gas temperatures from the height of the continuous flames to the height of the ceiling. This routine complements the *Ceiling Jet Temperature* procedure.

Theory The equation used in this procedure was developed by Alpert and Ward [31] and may be used to estimate the damages caused by the hot plume gases. Figure 4 in *Ceiling Jet Temperature*, section 5.5, illustrates some of the plume geometry variables used in eq (1).

$$T = T_\infty + 22.2 \frac{(K \dot{Q}_{fire})^{\frac{2}{3}}}{z^{\frac{5}{3}}} \quad ; \quad for \ \frac{r}{z} \leq 0.2 \tag{1}$$

$$t = \frac{20.3\ A_{ceiling}}{\dot{Q}^{\frac{1}{3}} z^{\frac{2}{3}}} \qquad (2)$$

A	Area (m²)
D_{fire}	Fire diameter (m)
K	Entrainment factor (1 for axisymmetric, 2 for wall-fire, 4 for corner fire)
Q	Total theoretical fire heat release rate (kW)
r	Radial distance from the center of the fire to the point of interest (m)
z	Entrainment height: vertical distance between the ceiling and the lowest point of the burning fuel (m)
t	Time (second)
T	Jet temperature at height, z, and radial distance, r, from the fire (°C)
T_∞	Ambient temperature (°C)

Notes The total theoretical, not the actual, heat release rate should be used to describe the fire in this and other ceiling/plume correlations by Alpert [31]. The total theoretical heat release rate may be obtained by multiplying the mass pyrolysis rate with the theoretical heat of combustion. Mass pyrolysis rates can be obtained through experimental measurement using load cells. Theoretical heats of combustion are available from handbooks. This fire specification is documented per Alpert's work.

The heat release rate of the fire is simulated as steady-state. The fire is modeled as a point source; no line fires are considered.

Radial locations from the vertical axis of the fire should be less than 0.2 times the height of the plume.

Radial location should be no further from the vertical axis of the fire than the shortest dimension of the compartment.

Temperature is conservative on the high side compared with experiments used to develop this correlation [31].

The continuous flame height is located at the base of the intermittent flaming region. The temperature of the continuous flaming region is about 800 °C in buoyancy-dominated diffusion flames, but may be as hot as the adiabatic flame temperature under ideal conditions (approximately 1800 °C). The height of the continuous flaming region is below the mean flame height. The mean flame height is defined as the elevation where flames appear 50% of the time [32], [33].

The temperature predictions from wall and corner configurations are theoretical.

The fire must be close enough to a wall that flames touch the surface before the user decides to choose the *Fire near a wall or corner* option. Combustible walls are not considered in this routine.

Standard pressure (101,325 Pa) conditions are used.

Time to hot layer development can be less than predicted for values of $z/D_{fire} \gg 1$.

A hot layer of gas has not developed at the ceiling.

Execution Select the compartment to be modeled from the fire scenario overview window, or select *Tools* from the desktop menu. If a compartment is selected from the fire scenario overview window, click on the tools icon. Select *Ceiling Plume Temperature* from the tools menu.

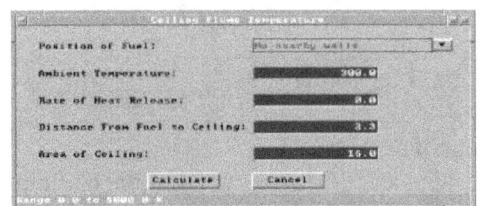

Note the display of range and measurement units at the bottom of the window as each edit widget is made the focus widget. To customize the units see the Advanced Features section 6.1.

Position of the Fuel: Select either *No nearby walls*, *Fuel package near a wall*, or *Fuel package in a corner*. Experiments have shown [30] that in order for reflection to apply, the fire flames must **touch** the wall.

Ambient Temperature: Compartment temperature at pre-fire conditions. If this procedure is selected for a compartment on the fire scenario overview window, the internal ambient temperature is provided as a default value.

Rate of Heat Release: Total theoretical fire heat release rate.

Distance From Fuel to Ceiling: Elevation difference between the lowest height where flames exist and where air can freely be entrained into the fire, and the height of the ceiling. If this procedure is selected for a compartment on the fire scenario overview window, a height corresponding to the ceiling of the selected compartment is provided as a default value.

Area of Ceiling: If this procedure is selected for a compartment on the fire scenario overview window, an area corresponding to the product of the compartment depth and width is provided as a default value.

Press the *Calculate* button to begin calculation. Upon completion, a results window is displayed.

The plume temperature and the time when a layer of hot gas that develops under the ceiling can interfere with the "no layer" assumption are reported.

5.7 Lateral Flame Spread

This procedure estimates the lateral spread of an attached flame along the surface of a thermally thick fuel. "Wind-aided" flame spread is an inappropriate application of this procedure. The procedure is appropriate for flame spread in a direction that is opposite, or normal to, the direction of the propagating flame front.

Theory The equations used in this procedure were developed by Quintiere and Harkelroad [34]. The material properties required by these equations may be experimentally obtained from the Lateral Ignition and Flame spread Test (LIFT) apparatus using procedures outlined in the above reference. The properties include the fuel flame spread parameter (ϕ), the fuel piloted-ignition temperature ($T_{ignition, pilot}$), and the fuel thermal inertia ($k\rho c_v$).

The equation for lateral flame spread appears in eq (1). The piloted-ignition temperature for many fuels is quite similar, as the software reference provided with this routine demonstrates.

$$v_{flame, lateral} = \frac{\phi}{k\rho c_v} \cdot \frac{1}{(T_{ig} - T_{surface})^2} \tag{1}$$

$v_{flame, lateral}$	Lateral rate of attached flame spread (m/s or ft/s)
ϕ	Ignition factor from flame spread test data (kW^2/m^3)
$k\rho c_v$	Fuel thermal inertia at flame preheat conditions ($kW^2 \cdot s/(m^4 \cdot K^2)$)
$T_{ignition, pilot}$	Piloted fuel ignition temperature (°C or °F)
$T_{surface}$	Unignited, ambient-surface temperature (°C or °F)

Notes Because temperature is raised to the second power, its impact on the estimated flame velocity is relatively large.

Thermal inertia ($k\rho c_v$) of the fuel is best measured as a single value and at elevated temperatures that accurately simulate conditions during actual flame spread.

It is strongly recommended that data for ϕ and $k\rho c_v$ be obtained from a single test per the suggestions of Quintiere and Harkelroad.

An inappropriate use of this model is for upward flame spread on a wall where unignited fuel in the "shadow" of the flame sheet receives significant preheating.

Experiments have shown correlation with this procedure for lateral extension on a horizontal fuel surface and downward flame extension on a vertical fuel surface [34], [35].

This procedure may not be appropriate for vertically-oriented fuel surfaces that drip when burning.

Execution Select the compartment to be modeled from the fire scenario overview window, or select *Tools* from the desktop menu. If a compartment is selected from the fire scenario overview window, click on the tools icon. Select *Lateral Flame Spread* from the tools menu.

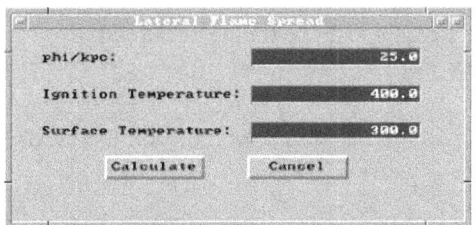

Note the display of range and measurement units at the bottom of the window as each edit widget is made the focus widget. To customize the units see the Advanced Features section 6.1.

phi/kpc: Calculated value of ϕ divided by $k\rho c_v$ for material of interest.

Ignition Temperature: $T_{ignition, pilot}$. If this procedure is selected for a compartment on the fire scenario overview window, the internal ambient temperature is provided as a default value.

Surface Temperature: $T_{surface}$. If this procedure is selected for a compartment on the fire scenario overview window, the internal ambient temperature is provided as a default value.

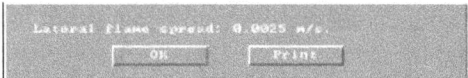

Press the *Calculate* button to begin calculation. Upon completion, a results window is displayed.

The lateral attached flame spread rate is reported.

5.8 Law's Severity Correlation

This procedure provides a systematic method whereby two fires, one a "real" fire and the other a standard, time-temperature fire, may be compared for equivalence regarding the structural damage imposed by each. Effectively, this procedure answers the question, "Given a known fire, what standard-fire resistance is needed to protect insulated structural members?" The standard, time-temperature fire follows the European specification [36], a close approximation to the ASTM E-119 standard time-temperature curve. The breadth of data from which this correlation was developed [37]. [38] as well as the length of time over which the correlation has been successfully used testifies to its robustness.

Theory The point of comparison between the "real" fire and the "standard" fire is the point in time that a critical temperature is achieved on the surface of a thermally-thick insulated structural element. Law chose 550 °C (1220 °F) as the critical temperature because at this temperature steel's modulus of elasticity or strength is dramatically reduced. However, other critical temperatures could have been chosen just as readily, the net result being a relationship bearing the same form as eq (1) with different constants. It has been demonstrated that other thermally-thick-insulated elements, *e.g.*, concrete and heavy-timber, could be analyzed with this method. Although these other insulated-elements may not fail at the "critical" temperature assumed with this procedure, these elements would nonetheless experience a similar surface temperature of

$$t_{effective\,resistance} = \frac{m_{fuel}}{\sqrt{A_{vent} A_{room}}} \qquad (1)$$

approximately 550 °C.

$t_{effective}$ — Duration of exposure to the ISO standard time-temperature fire-resistance test that is correlationally equivalent to the "real" fire exposure. (min)

A_{vent} — Effective area of all vents. When more than one vent exists, use the method outlined in section 5.13 for calculating an equivalent area. (m²)

A_{room} — Area of the compartment surfaces. The calculations *do* include the floor and ceiling area, as well as the area of the walls, less A_{vent}. (m²)

m_{fuel} — Mass of dry wood burning in a crib configuration that is equivalent to the total energy released by the "real" fire. (kg)

Notes This procedure requires as input a mass of wood fuel (m_{fuel}) releasing an equivalent energy upon completion of burning as released from complete burning of the "actual" fuel. The mass

$$m_{wood} = \frac{m_{fuel,actual} \Delta H_{c_{fuel,actual}}}{\Delta H_{c_{wood}}} \qquad (2)$$

of wood fuel needed as input for the **Fire Load** input parameter described in program execution section may be determined from eq (2). Informed and/or engineering judgement is needed to determine what are appropriate and prudent values for the heats of combustion used in eq (2).

The concept that an exposure time in a standard time-temperature fire can be related to a different exposure time in an actual fire hinges upon the premise that a thermally-protected steel column within each exposure attains the same temperature at the completion of each fire.

The correlation assumes the materials providing the estimated fire resistance protection are thermally thick. Thermally thin protection or exposed steel members may not be a valid application for this correlation [37].

The tests used for development of the correlation were conducted in small- and full-scale compartments with concrete and fibre-board insulated walls and a variety of ventilation sizes. Various fuels including tires, liquids, wood cribs, and furniture were used in correlating eq (1) [37,38].

The test configurations all contained open vents. It is recommended that the compartment being investigated possess at least a 0.4 m² (4 ft²) opening. This recommendation is based upon engineering judgement [39].

Execution Select the compartment to be modeled from the fire scenario overview window, or select *Tools*

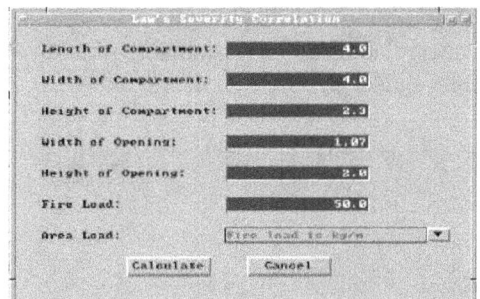

from the desktop menu. If a compartment is selected from the fire scenario overview window, click on the tools icon. Select *Law's Severity Correlation* from the tools menu. The following window is displayed:

Note the display of range and measurement units at the bottom of the window as each edit widget is made the focus widget. To customize the units see the Advanced Features section 6.1.

Length of the compartment: Depth of the compartment as measured forward from the left, rear corner of the compartment. If this procedure is selected for a compartment on the fire scenario overview window, the depth of the current compartment is displayed as a default value.

Width of the compartment: Width of the compartment as measured across from the left, rear corner of the compartment. If this procedure is selected for a compartment on the fire scenario overview window, the width of the current compartment is provided as a default value.

Height of the compartment: Height of the compartment. If this procedure is selected for a compartment on the fire scenario overview window, the height of the current compartment is provided as a default value.

Width of the opening: Width of the opening or width of an equivalent vent opening calculated from eq (3) in section 5.13 when representing multiple vents in the compartment. If this procedure is selected for a compartment on the fire scenario overview window, the width of a virtual vent that has an area equivalent (for the purposes of determining flashover) to the combined area of all individual vents in the compartment is provided as a default value.

Height of the opening: Height of the opening or height of an equivalent vent opening calculated from eq (3) in section 5.13 when representing multiple vents in the compartment. If this procedure is selected for a compartment on the fire scenario overview window, the height of an equivalent vent using the difference between the elevation of the highest point and the lowest point among all of the vents in the compartment is provided as a default value.

Fire Load: Mass of wood producing a fire equal to the actual fire (kg or lb_m).

Area Load: Toggle selection between *Fire Load is kg/m²* and *Fire Load is gross load*. Selection setting determines units associated with input for the fire load input line.

Press the *Calculate* button to begin calculation. Upon completion, a results window is displayed.

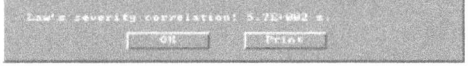

The exposure time in the standard fire test that is equivalent to the time of exposure in a "real" fire is reported.

5.9 Mass Flow Through a Vent

This procedure uses an iterative process to determine an approximate solution for mass flow into and out of a single, naturally-ventilated opening to a compartment containing a steady-state fire. This routine is similar but not identical to the *Smoke Flow Through an Opening* estimation routine.

Theory Conservation of mass is used to solve numerically the gas-mass flow rate into a naturally ventilated compartment with steady-state elevated temperatures, see eq (1).

$$\{Generation\ Rate\} + \dot{m}_{in} - \dot{m}_{out} = \frac{d\,m_{cv}}{dt} \tag{1}$$

$$z_{neutral\ plane} = \frac{1}{1 + \left(\frac{T}{T_\infty}\right)^{\frac{1}{3}} \left(1 + \frac{\dot{m}_{pyrolysis}}{\dot{m}_{in}}\right)^{\frac{2}{3}}} \tag{2}$$

$$M_o = \frac{\sqrt{\theta}}{1+\theta}\left(1 - z_{neutral\,plane}\right)^{3/2}, \quad \text{where } \theta = \frac{T-T_\infty}{T_\infty} \tag{3}$$

$$\dot{m}_{out} = \frac{2}{3} C M_o A_{vent} \rho_\infty \sqrt{2 g z_{vent}} \qquad (4)$$

The mass flow calculation is solved using the conservation of mass principal whereas the smoke flow calculation in section 5.12 is solved using Bernoulli flow and orifice equation assumptions. The solution process begins with the user-specified fuel pyrolysis rate from the "generation rate" term in eq (1). The procedure then estimates a mass inflow rate based upon door width. This inflow rate is used to calculate the neutral-plane elevation ($z_{neutral-plane}$) using eq (2). The neutral plane elevation is subsequently used to calculate the mass outflow rate, see eq (3) and (4). If mass conservation does not close within the specified criteria, then the bisection numerical technique estimates again and the process repeats until convergence or excess iterations are achieved [40]. Some of the first work on these vent flow rates was presented by Kawagoe [41].

In the conservation of mass equation, the pyrolysis rate is a source term, and as such its value is inserted in the generation rate term of eq (1). Since this procedure assumes a steady-state elevated temperature in the compartment, the net rate-of-change in mass within the compartment ($d[m_{cv}]/dt$) is zero.

$\dot{m}_{pyrolysis}$	Mass generation rate of the fuel (g/s)
$\dot{m}_{cv}$	Net mass of gas within the control volume (the control volume is the compartment air-volume not including the wall/ceiling materials).
z_{vent}	Height of the vent opening from soffit to sill (m)
$z_{neutral\ plane}$	Height of the neutral plane in the vertical opening of the door (m)
T	Temperature of the hot gas layer (K)
T_∞	Temperature of the ambient, outside air (K)
θ	Non-dimensionalized temperature variable $(T - T_\infty)/T_\infty$

Notes Air flows are motivated by buoyancy forces only: no mechanical pressurization, stack effect or wind effects are considered.

Conditions are steady state: pyrolysis rate is constant, layer temperatures are constant, layer heat losses are constant, momentum flow across the vent is constant. Examples of steady-state conditions could be flashover, or fuel controlled burning when the fire is not growing.

The gas in the compartment is either at a uniform temperature, *e.g.*, flashover, or is in a ventilation-limited condition. This procedure is inappropriate for early stages in a fire when a hot layer increases in thickness and the expansion of gases causes only an outflow from the compartment.

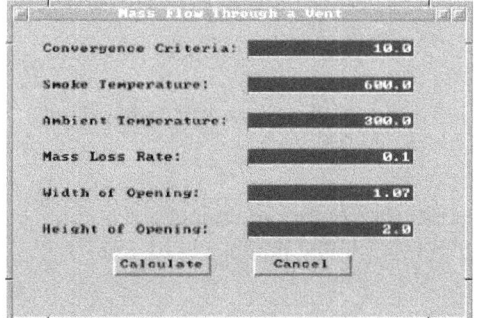

There is either one opening in the space, or all of the openings are at approximately the same level in the compartment. If openings must be combined, they should be combined per the method used in Thomas's Flashover Correlation, section 5.13.

Execution Select the compartment to be modeled from the fire scenario overview window, or select *Tools* from the desktop menu. If a compartment is selected from the fire scenario overview window, click on the tools icon. Select *Mass Flow Through a Vent* from the tools menu.

Note the display of range and measurement units at the bottom of the window as each edit widget is made the focus widget. To customize the units see the Advanced Features section 6.1.

Convergence Criteria: Criterion for numerical solution of eq (2).

Smoke Temperature: Temperature of the hot, upper layer of gas. If this procedure is selected for a compartment on the fire scenario overview window, the previously specified internal ambient temperature is provided as a default value.

Ambient Temperature: Temperature of the ambient, outside temperature. If this procedure is selected for a compartment on the fire scenario overview window, the previously specified external ambient temperature is provided as a default value.

Mass Loss Rate: Fuel pyrolysis rate.

Width of the opening: Width of the opening or width of an equivalent vent opening calculated from eq (3) in section 5.13 when representing multiple vents in the compartment. If this procedure is selected for a compartment on the fire scenario overview window, the width of a virtual vent that has an area equivalent (for the purposes of determining flashover) to the combined area of all individual vents in the compartment is provided as a default value.

Height of the opening: Height of the opening or height of an equivalent vent opening calculated from eq (3) in section 5.13 when representing multiple vents in the compartment. If this procedure is selected for a compartment on the fire scenario overview window, the height of an equivalent vent using the difference between the elevation of the highest point and the lowest point among all of the vents in the compartment is provided as a default value.

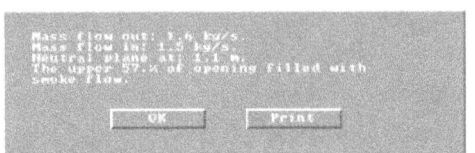

Execution: Press the *Calculate* button to begin calculation. Upon completion, a results window is displayed.

The mass flow rates in and out of the compartment are reported. In addition, the absolute height of the neutral plane above the bottom of the vent is displayed. This variable is also expressed as the nondimensionsal height of the neutral plane in the door. Finally, a percentage of vent area filled is calculated and displayed.

5.10 Plume Filling Rate

This procedure estimates the volume flow of smoke and entrained air in a plume at a point above the flames of a fire of constant heat release rate.

Theory There are currently several models [42], [43], [44], [45] that estimate entrainment into a rising buoyant plume. Each model provides roughly the same accuracy with no individual model clearly outperforming the others in all cases. This procedure uses a model originated by Zukoski and later modified by Cooper and Stroup [46].

The equation used by this procedure is:

$$\dot{V}(z) = 0.0026\left(1-\chi_r\right)\dot{Q} + 0.006047\left(\left(1-\chi_r-\chi_o\right)\dot{Q}\right)^{1/3} z^{5/3} \qquad (1)$$

$\dot{V}(z)$	Volumetric flow rate of all gases in the plume at height z (Liters/s)
$\dot{Q}$	Theoretical fire heat release rate (kW/s)
χ_r	Fraction of Q released through radiative heat transfer
χ_o	Fraction of Q not released via radiative or convective heat into the plume
z	Height in the plume where V(z) is calculated (m)

Notes This procedure applies to steady-state fires.

The input parameter χ_o may be used to account for combustion inefficiencies and/or heat from the fire that is expended in pyrolyzing fuel. If χ_o is non-zero, then the sum of χ_o and χ_r should be less than one.

This procedure should not be applied at heights equal to or less than the mean flame height. The mean flame height is that elevation on the central fire axis where flames appear 50% of the time. The mean flame height also correlates with an average gas temperature of 500 °C. [43].

Other plume models have been verified in large atria [47].

The approximate conversion between Liters/s and cubic feet per minute is 2.12. To convert from L/s to cfm, multiply the number representing flow in L/s by 2.12.

The fire is considered a point source; i.e., no line fires or fire areas in a distributed sense are considered.

This procedure does not consider wall or corner fires.

The buoyant gas has no appreciable horizontal momentum.

There is no contact between the walls of the compartment and the plume (mathematically the walls are a distance, r, from the fire: r ≥ 0.2 * H).

This procedure should not be used on predominantly momentum-driven plumes.

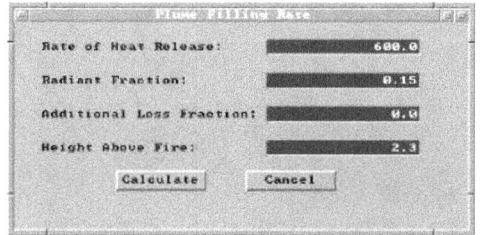

The plume is not tilted from the vertical and the plume is not experiencing wind- or mechanically-aided entrainment.

Execution Select the compartment to be modeled from the fire scenario overview window, or select *Tools* from the desktop menu. If a compartment is selected from the fire scenario overview window, click on the tools icon. Select *Plume Filling Rate* from the tools menu.

Note the display of range and measurement units at the bottom of the window as each edit widget is made the focus widget. To customize the units see the Advanced Features section 6.1.

Rate of Heat Release: Fire heat release rate.

Radiant Fraction: Fraction of heat distributed via radiative energy [48].

Additional Loss Fraction: Fraction of heat not convected and not radiatively distributed.

Height Above Fire: Elevation difference between the point of interest in the plume and the lowest height where entrainment begins. For diffusion flames this is usually the base of the flames.

Press the *Calculate* button to begin calculation. Upon completion, a results window is displayed.

Volumetric flow rate in the plume at the specified height above the fire is reported.

5.11 Radiant Ignition of a Near Fuel

This is a quick, simplistic estimation method for determining what size fire will radiatively ignite a nearby fuel; no flame impingement is assumed.

Theory The equations in this procedure were obtained through correlation of experimental data [49]. The experiments examined the fire sizes necessary to ignite a second, remote, initially non-burning fuel item that was not in direct contact with the flame or convective flow of the original fire. The experimental data provided a correlation between the peak heat release rate of the first-burning item and the maximum distance to a second non-burning fuel item that would result in ignition. Babrauskas found the second item could usually be categorized into one of three groups--based upon material and size:

- **Easily Ignited** - material ignites when it receives a radiant flux of 10 kW/m² or greater. Examples are thin materials such as curtains or draperies.

$$\dot{Q}_{fire} = 30.0 \times 10^{(\frac{Distance + 0.08}{0.89})} \tag{1}$$

- **Normally Resistant to Ignition** - material ignites when it receives a radiant flux of 20 kW/m² or greater. Examples are upholstered furniture and other materials with significant mass but small thermal inertia ($k\rho c_p$).

$$\dot{Q}_{fire} = 30(\frac{Distance + 0.05}{0.019}) \tag{2}$$

- **Difficult to Ignite** - material ignites when it receives a radiant flux of 40 kW/m² or greater. Examples are thermoset plastics and other thick materials (greater than 0.013 m [½ in.]) with substantial thermal inertia (kρc$_p$).

$$\dot{Q}_{fire} = 30(\frac{Distance + 0.02}{0.0092}) \quad (3)$$

Q$_{fire}$ Fire heat release rate (kW)
α   Regression coefficient of experimentally measured data (m)
β   Regression coefficient of experimentally measured data (m)

Notes The least accuracy is obtained when analyzing "Easy to ignite" items. This is because the required heat release rates for igniting these types of fuels are low enough that small changes in fuel properties can result in large percentage changes for the required ignition heat release rates.

Target fuels do not have flame impingement considered in the ignition analysis.

The distance between the exposed fuel item and the initial burning fuel is small enough to nullify a point source radiation assumption. The target-fuel item assumedly sees a broad fire such as that produced by a free-standing upholstered chair or side of a couch.

Fuel types included in the experimental correlation were: wood, plywood, plywood laminates, paper, polyurethane and polyethylene. These fuel types have a radiative fraction varying from 0.6 for polyurethane to 0.3 for wood pine [50].

Execution Select the compartment to be modeled from the fire scenario overview window, or select *Tools* from the desktop menu. If a compartment is selected from the fire scenario overview window, click on the tools icon. Select *Radiant Ignition* from the tools menu.

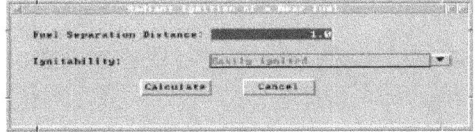

Note the display of range and measurement units at the bottom of the window as each edit widget is made the focus widget. To customize the units see the Advanced Features section 6.1.

Fuel Separation Distance: Separation of fuel packages.

Ignitability: Select either *Easily Ignited, Normally Resistant to Ignition*, or *Difficult to Ignite*. For easily ignitable fuels, the energy flux is 10 kW/m². For normal fuels, the energy flux is 20 Kw/m², and for difficult to ignite fuels, the energy flux is 40 kW/m².

Press the *Calculate* button to begin calculation. Upon completion, a results window is displayed.

The fire size of the initially burning fuel needed to ignite the second fuel source is reported.

5.12 Smoke Flow Through an Opening

This procedure estimates the steady-state volumetric flow rate of heated gas at elevated temperatures through an opening in an enclosure. It is appropriate for measuring post-flashover or steady-state smoke leakage through open doors or cracks around closed doors. Whereas this procedure solves for the volumetric smoke flow rate at a given temperature, the *Mass Flow Through a Vent* discussed in section 5.13 solves for the vent flow in terms of mass.

Theory The theory for smoke flow through an opening due to buoyancy forces is developed in *Design of Smoke Control Management Systems*, [51]. The derivation foundation is the classical orifice eq (1). In contrast, the mass flow through a vent calculation discussed previously in section 5.13 iterates on a solution satisfying mass conservation. The velocity term in eq (1) may be substituted for by rearranging the Bernoulli expression, eq (2), and solving for velocity in terms of the other parameters, eq (2b). The assumptions for Bernoulli flow are presented in the following notes section. The pressure term in eq (2b) is solved by rearranging the ideal gas law, see eq (3), and expressing pressure in terms of gas density.

$$\dot{V} = CA_{vent} v \tag{1}$$

$$\frac{\Delta P_1}{\rho_1} + \frac{v_1^2}{2} + gz_1 = \frac{\Delta P_2}{\rho_2} + \frac{v_2^2}{2} + gz_2 \tag{2}$$

$$v_2 = \sqrt{\frac{\Delta P_{(1-2)}}{\rho_2}} \tag{2b}$$

$$\Delta P_{1-2} = \Delta \rho_{1-2} g h \tag{3}$$

$$\Delta \rho_{1-2} = \frac{P(MW)}{R} \left| \frac{1}{T_1} - \frac{1}{T_2} \right| \tag{3b}$$

$$\dot{V} = CA_{vent} \sqrt{\frac{2ghP(MW)}{\rho_2 R} \left| \frac{1}{T_1} - \frac{1}{T_2} \right|} \tag{4}$$

A_{vent}	Area of the vent that allows smoke movement (m²)
C	Orifice coefficient (0.8)
g	Acceleration constant equal to Earth's surface gravity (9.81 m/s²)
h	Height of the neutral plane (m)
MW	Ambient fluid molecular weight (28.95·10⁻³ kg air/(g·mole))
P	Standard pressure (101,325 Pa)
ΔP_{1-2}	Pressure difference across the vent (Pa)
R	Universal gas constant (8.314 J/((g·mole)·K)))
T	Air temperature (K)
T_∞	Ambient air temperature (294 K)

V̇	Volumetric vent flow rate (m³/s)
z	Elevation difference (m)
v	Smoke velocity (m/s)
Δρ$_{1-2}$	Density difference across the vent (kg/m³)

Subscript

1	Location inside the compartment
2	Location just beyond the vent

Notes Assumes uniform depth of smoke in the vent area.

Steady-state flow conditions are assumed.

Bernoulli fluid is steady-state, incompressible, and nonviscous.

Only buoyancy driven smoke flow is considered.

Standard pressure, no stack effect, no air-handling systems, no wind forces, no unrelieved pressure-volume work by the expanding, hot smoke.

Standard gravitational acceleration at Earth's surface, *i.e.*, smoke flow is not considered in systems undergoing additional accelerations. The heated smoke layer is quiescent away from the vent.

Inappropriate for duct-like openings where the passage length is significantly greater than the narrow dimension of the opening.

Air is the ambient gas (or any other ambient gas having a similar molecular weight) at a temperature of 21 °C (70 °F).

The temperature should characterize the average conditions throughout the smoke layer.

Execution Select the compartment to be modeled from the fire scenario overview window, or select *Tools* from the desktop menu. If a compartment is selected from the fire scenario overview window, click on the tools icon. Select *Smoke Flow Through an Opening* from the tools menu.

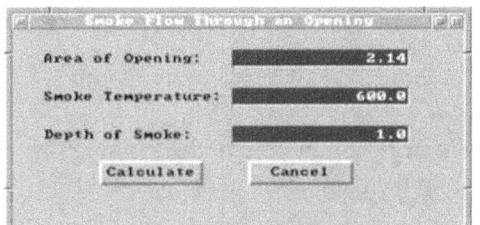

Note the display of range and measurement units at the bottom of the window as each edit widget is made the focus widget. To customize the units see the Advanced Features section 6.1.

Area of Opening: Area of the opening or area of an equivalent vent opening calculated from eq (3) in section 5.13 when representing multiple vents in the compartment. If this procedure is selected for a compartment on the fire scenario overview window, the area of a virtual vent as discussed in the Thomas's Flashover Correlation is provided as a default value.

Smoke Temperature: Temperature of the hot, upper layer of gas. If this procedure is selected for a compartment on the fire scenario overview window, the previously specified internal ambient temperature is provided as a default value.

Depth of Smoke: Depth of the smoke layer.

Press the *Calculate* button to begin calculation. Upon completion, a results window is displayed.

 The volumetric smoke flow rate out of the vent area is reported.

5.13 Thomas's Flashover Correlation

This procedure quickly estimates the amount of energy needed to produce flashover in a compartment.

Theory This procedure [52] results from simplifications applied to a hot-layer energy balance on a compartment with a fire. These simplifications resulted in eq (1). The first term represents heat losses to the "...total internal surface area of the compartment...", and the second term represents energy flow out of the vent opening. The two constants in eq (1) represent values correlated to experimental flashover conditions.

$$\dot{Q} = 7.8 A_{room} + 378 (A_{vent}\sqrt{H_{vent}})_{equivalent} \tag{1}$$

$$A_{room} = A_{floor} + A_{ceiling} + A_{walls} - A_{vents_{equivalent}} \tag{2}$$

$$A_{vents_{equivalent}} = H_{vent_{equivalent}} \cdot W_{vent_{equivalent}} \tag{3}$$

$$W_{vent_{equivalent}} = \frac{(A_{vent}\sqrt{H_{vent}})_1 + (A_{vent}\sqrt{H_{vent}})_2 + \ldots}{H_{vent_{equivalent}}^{\frac{3}{2}}} \tag{4}$$

Q = Fire heat release rate (kW)
A_{vent} = Area of the vent (m²)
$H_{vent, equivalent}$ = The difference between the elevation of the highest point among all of the vents and the lowest point among all of the vents (m).
$W_{vent, equivalent}$ = The width of a virtual vent that has an area equivalent (for the purposes of determining flashover) to the combined area of all individual vents from the compartment of consideration (m).

Notes The formulation of the energy balance considered heat losses from the hot gas layer and heated walls to the cooler lower walls and floor surfaces. The term A_{room} should include all surfaces inside the compartment, exclusive of the vent area.

The fire area should not be subtracted from the floor area as the fire will conduct and convect heat into the floor underneath the fuel footprint.

The equation does not know where the vent is located, nor whether the vent is a window or a door; however, the equation was developed from tests that included window venting.

The equation does not consider whether the walls are insulated or not. Use of the equation for compartments with thin metal walls may therefore be inappropriate. The experiments included compartments with thermally thick walls and fires of wood cribs. The equation was later verified in gypsum lined compartments with furniture fires [53].

Verification with fast growing fires: the correlation was developed from fast not slow growing fires.

This procedure was correlated from experiments conducted in compartments not exceeding 16 m² in floor area.

The equation predicts flashover in spaces without ventilation. This prediction is unlikely due to oxygen starvation of the fire.

Execution Select the compartment to be modeled from the fire scenario overview window, or select *Tools* from the desktop menu. If a compartment is selected from the fire scenario overview window, click on the tools icon. Select *Thomas Flashover* from the tools menu.

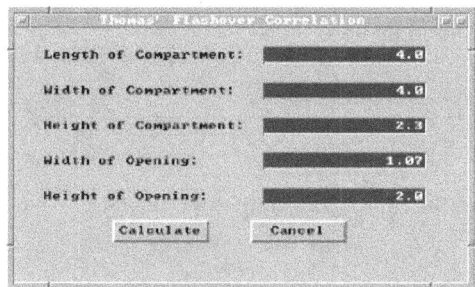

Note the display of range and measurement units at the bottom of the window as each edit widget is made the focus widget. To customize the units see the Advanced Features section 6.1.

Length of the compartment: Depth of the compartment as measured forward from the left, rear corner of the compartment. If this procedure is selected for a compartment on the fire scenario overview window, the depth of the current compartment is displayed as a default value.

Width of the compartment: Width of the compartment as measured across from the left, rear corner of the compartment. If this procedure is selected for a compartment on the fire scenario overview window, the width of the current compartment is provided as a default value.

Height of the compartment: Height of the compartment. If this procedure is selected for a compartment on the fire scenario overview window, the height of the current compartment is provided as a default value.

Width of the opening: Width of the opening or width of an equivalent vent opening calculated from eq (3) when representing multiple vents in the compartment. If this procedure is selected for a compartment on the fire scenario overview window, the width of a virtual vent that has an area equivalent (for the purposes of determining flashover) to the combined area of all individual vents in the compartment is provided as a default value.

Height of the opening: Height of the opening or height of an equivalent vent opening calculated from eq (3) when representing multiple vents in the compartment. If this procedure is selected for a compartment on the fire scenario overview window, the height of an equivalent vent using the difference between the elevation of the highest point and the lowest point among all of the vents in the compartment is provided as a default value.

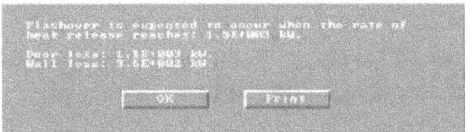

Press the *Calculate* button to begin calculation. Upon completion, a results window is displayed.

The estimated fire heat release rate that will create flashover is displayed. In addition, the estimated energy losses from gas flow out of the door and from the gas to compartment wall surfaces are reported.

5.14 Ventilation Limit

This procedure estimates the maximum post-flashover fire size sustainable in a compartment based upon the ventilation geometry. Vent geometry can control the fire size by limiting the amount of air entering the compartment and hence limiting the amount of oxygen that may combine with the fuel.

Theory Kawagoe [54] originally presented the idea that fire heat release rate within a compartment could be limited by ventilation geometry. This idea was borne out by Kawagoe's original--and many subsequent post-flashover experiments (for example, [55], [56]). The equation implemented in this procedure is presented in eq (1) where the mass flow rate of air into the compartment ½$A_o H_o^{½}$ is in kg/s.

$$\dot{Q}_{VL} = \chi_A \Delta H_{c,air} \frac{1}{2} A_o \sqrt{H_o} \qquad (1)$$

Q_{VL}	Limit of fire heat release rate supportable by a naturally ventilated compartment (kW)
χ_A	Combustion efficiency
A_o	Area of the opening (m²)
H_o	Height of opening (m)
$\Delta h_{c,air}$	Fuel heat of combustion per kilogram of air that oxidizes fuel (~3000 kJ/kg).

Notes It is possible to calculate the dimensions of a single vent that will sustain the fire burning rate allowed by several individual vents each contributing air (oxidizer) to the fire up to the limit supported by their geometrical size. The dimensions of this equivalent vent are obtainable from eq (3) in section 5.13. If the ventilation limit procedure is selected for an existing compartment in the input editor, the equivalent vent dimensions for that compartment are provided as default inputs.

The equivalent vent dimension approach is not appropriate for use when vents are located at significantly different elevations in the wall.

This routine is not applicable to the early times in the growth of a compartment fire when fuel-limited burning occurs. In this situation, more than enough air needed to sustain burning passes through the vent and reaches the fuel.

This routine will calculate the heat released inside the compartment; however, it is possible that additional heat may be released outside of the compartment that is unaccounted for by this procedure. This can occur if during ventilation limited burning, the fire pyrolyzes more fuel than the air is capable of burning. The unburned pyrolyzate will be carried out the vent and may burn in a "door or window jet" providing that the pyrolyzate concentrations are high enough, hot enough, and sufficient oxygen for combustion is present.

Assuming χ_A as unity results in a prediction for the largest possible ventilation limit fire; this may be appropriate for design fires used in life-safety hazard analysis.

Execution Select the compartment to be modeled from the fire scenario overview window, or select *Tools* from the desktop menu. If a compartment is selected from the fire scenario overview window, click on the tools icon. Select *Ventilation Limit* from the tools menu displayed.

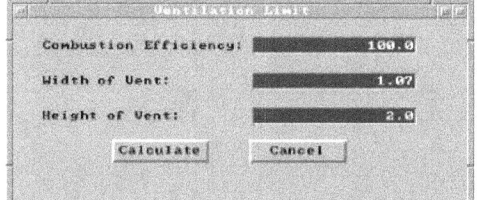

Note the display of range and measurement units at the bottom of the window as each edit widget is made the focus widget. To customize the units see the Advanced Features section 6.1.

Combustion efficiency: For combustion efficiency, refer to reference below.

Width of vent: Width of the vent or width of an equivalent vent calculated from eq (3) in section 5.13 when representing multiple vents in the compartment. If this procedure is selected for a compartment on the fire scenario overview window, the width of a virtual vent that has an area equivalent (for the purposes of determining flashover) to the combined area of all individual vents in the compartment is provided as a default value.

Height of vent: Height of the vent or height of an equivalent vent calculated from eq (3) in section 5.13 when representing multiple vents in the compartment. If this procedure is selected for a compartment on the fire scenario overview window, the height of an equivalent vent using the difference between the elevation of the highest point and the lowest point among all of the vents in the compartment is provided as a default value.

Press the *Calculate* button to begin calculation. Upon completion, a results window is displayed.

The Fire heat release rate as limited by ventilation geometry is reported.

6 Advanced Features

Users unfamiliar with the concepts of a Graphical User Interface (GUI) are strongly encouraged to review these concepts in the Getting Started section of this guide before continuing with the Utilities section. Familiarity with GUI terminology and the use of a mouse is assumed throughout the remainder of this section.

6.1 Changing Display Units

As the user enters information for edit widgets within the GUI shell, range messages for the current widget are displayed at the bottom of each input window along with the user selected measurement units. A new installation of the FASTLite software displays measurements in standard SI units. Users can customize the display measurement units to appropriately reflect the environment in which they work by selecting *Options* from the desktop menu then selecting *User Specified Units*.

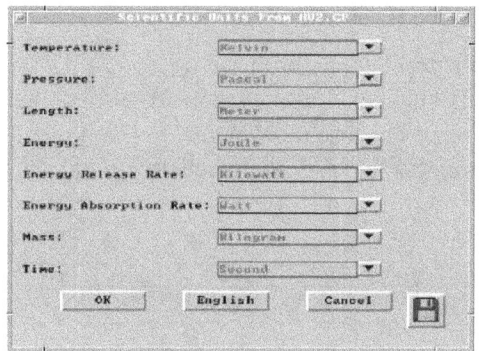

Modifications are made to a set of eight (8) base measurements from which all other measurements within the GUI shell are derived. These eight measurements include: temperature, pressure, length, energy, energy release rate, energy absorption rate, mass, and time. Assuming that the installation selections are in place, SI measurement display unit settings for each of these measurements are: temperature - K, pressure - Pa, energy - J, energy release rate - W, energy absorption rate - W, mass - kg, and time - s. With these base measurement settings, the display units for some derived measurements are: specific heat - J/kg*K, mass loss rate - kg/s, and volumetric rate change - m^3/s. To modify the settings for any of the base measurements, click  on the pull-down icon to the right of the measurement.

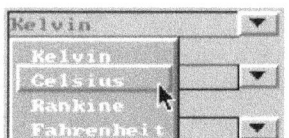

A pull-down menu is displayed, listing the available measurement units. Select the display units desired by clicking on the corresponding menu entry.

It is important to remember that modifications made are reflected in all measurements derived from the selected base measurement. For example, if units for length are changed from meters to feet, area references in the GUI shell will indicate ft^2, volume will indicate ft^3, and velocity will indicate ft/s.

To modify the default SI setting to use English units, click on the center text button at the bottom of the window. Units for all base measurements are set to standard English values. The text displayed on this button is dependent on the current setting of the base measurements. If current units are SI, the text *English* is displayed. If current units are English, the text *Metric* is displayed.

If the selected units are to be used every time the FASTLite software is run, click on the disk icon in the lower right corner of the window. This saves current settings to the configuration file indicated in both the window title and the first edit widget.

If the selected units are to be used only for the current FASTLite software run, click on the *OK* button and continue with input in the GUI shell.

6.2 Input Editor Error Log

As input files are read, and the overview windows in the input editor are created, a file of errors encountered  in the input file is generated by the FASTLite software. If critical errors are encountered, a red message window is displayed prior to displaying the fire scenario overview window. To view the list of errors, select *Utilities* from the desktop menu, then select *View Errors*.

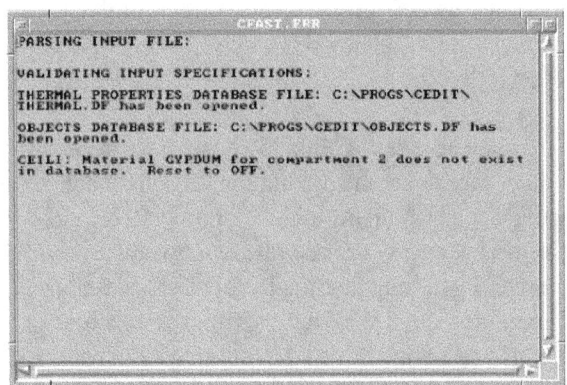

Use the vertical scroll bar to view other sections of the file. Refer to the GUI Terminology section 1.2.2 of this reference guide for an explanation of the use of scroll bars in the GUI interface.

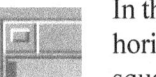 In the top left corner of the window, a horizontal bar is displayed inside a small square. To close the window, position the mouse pointer on this line and double-click.

6.3 Copy File

FASTLite supports a copy file utility to enable the user to copy one input file, database file, configuration file, or other ASCII text file to another file name prior to making modifications. To copy a file, select *Utilities* from the desktop menu, then select *Copy File*. A window is displayed.

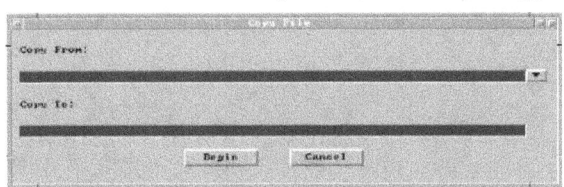

Copy From: Enter the full filename including path and directory to be copied. If a partial filename is known, enter the known characters along with * to indicate those parts of the filename for which characters are not known, press the pull-down icon. If the pull-down icon is pressed, the file name window is displayed.

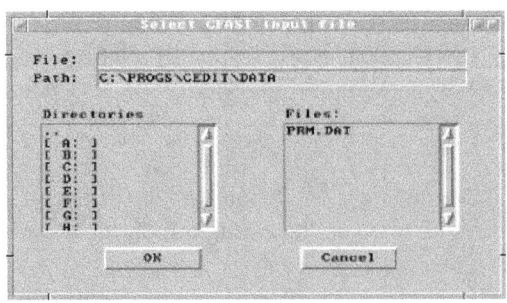

The file specification window consists of four key elements: the filename edit widget, the directory edit widget, the directory selection list, and the filename selection list. To search directories other than the current directory displayed in the directory edit widget, enter the new directory name in the directory edit widget and press Enter. Directories may also be selected by double-clicking on the drive or directory in the directory selection list. If the filename is known, enter it in the filename edit widget and press Enter. Do not include the drive or directory in the filename. Drive and directory specifications are handled only through the directory edit widget and the directory selection list widget. Once the desired file is found, double-click on the entry in the filename selection list, or single-click and press the *OK* button.

Copy To: Enter the full filename including path and directory where the file is to be copied. If the specified file already exists, a warning message is displayed requesting confirmation before overwriting the file.

Press *Begin* to begin copying the file, or *Cancel* to exit this procedure without copying the file.

6.4 Print File

FASTLite supports a print file utility to enable the user to print an input file or other ASCII text file through a selected printer port. To print a file, select *Utilities* from the desktop menu, then select *Print File*. The print file window is displayed.

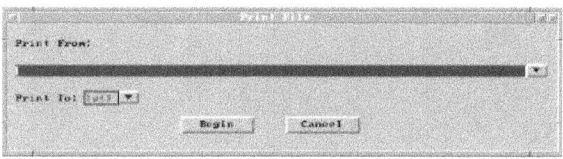

Print From: Enter the full filename including path and directory to be printed. If a partial filename is known, enter the known characters along with * to indicate those parts of the filename for which characters are not known, press the pull-down icon. If the pull-down icon is pressed, the file name window is displayed:

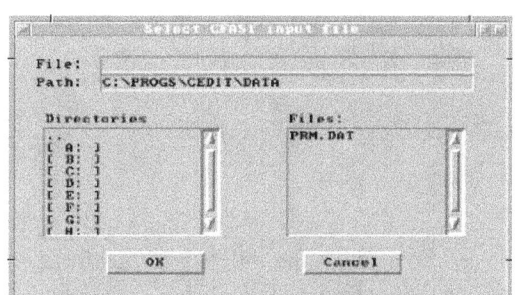

The file specification window consists of four key elements: the filename edit widget, the directory edit widget, the directory selection list, and the filename selection list. To search directories other than the current directory displayed in the directory edit widget, enter the new directory name in the directory edit widget and press Enter. Directories may also be selected by double-clicking on the drive or directory in the directory selection list. If the filename is known, enter it in the filename edit widget and press Enter. Do not include the drive or directory in the filename. Drive and directory specifications are handled only through the directory edit widget and the directory selection list widget. Once the desired file is found, double-click on the entry in the filename selection list, or single-click and press the *OK* button.

Print To: Selections include: *lpt1*, *lpt2*, *com1*, *com2*, *com3*, and *com4*. If the specified port is not available, a warning message is displayed requesting confirmation prior to printing the file.

Press *Begin* to begin printing the file, or *Cancel* to exit this procedure without printing the file.

6.5 View File

FASTLite supports a view file utility to enable the user to view an input file, configuration file, or other ASCII text file. To view a file, select *Utilities* from the desktop menu, then select *View File*. The view file window is displayed.

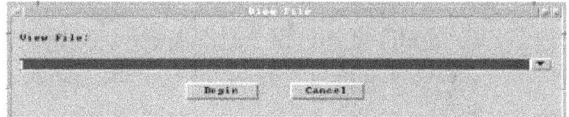

View File: Enter the full filename including path and directory to be viewed. If a partial filename is known, enter the known characters along with * to indicate those parts of the filename for which characters are not known, press the pull-down icon. If the pull-down icon is pressed, the following window is displayed:

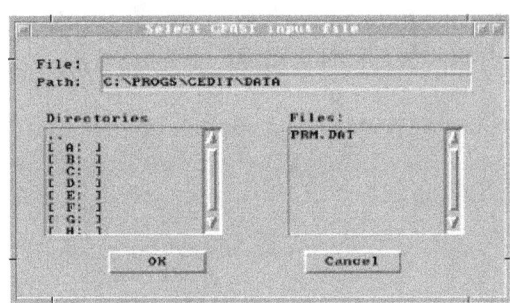

The file specification window consists of four key elements: the filename edit widget, the directory edit widget, the directory selection list, and the filename selection list. To search directories other than the current directory displayed in the directory edit widget, enter the new directory name in the directory edit widget and press Enter. Directories may also be selected by double-clicking on the drive or directory in the directory selection list. If the filename is known, enter it in the filename edit widget and press Enter. Do not include the drive or directory in the filename. Drive and directory specifications are handled only through the directory edit widget and the directory selection list widget. Once the desired file is found, double-click on the entry in the filename selection list, or single-click and press the *OK* button.

Press *Begin* to begin viewing the file, or *Cancel* to exit this procedure without viewing the file.

Use the vertical scroll bar to view other sections of the file. Refer to the GUI Terminology section 1.2.2 of this reference guide for an explanation of the use of scroll bars in the GUI interface.

 In the top left corner of the window, a horizontal bar is displayed inside a small square. To close the window, position the mouse pointer on this line and double-click.

6.6 Delete File

FASTLite supports a delete file utility to enable the user to delete an input file, configuration file, or other ASCII text file. To delete a file, select *Utilities* from the desktop menu, then select *Delete File*. The delete file window is displayed.

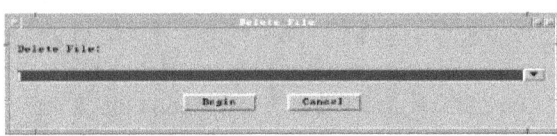

Delete File: Enter the full filename including path and directory to be deleted. If a partial filename is known, enter the known characters along with * to indicate those parts of the filename for which characters are not known, press the pull-down icon. If the pull-down icon is pressed, the file name window is displayed.

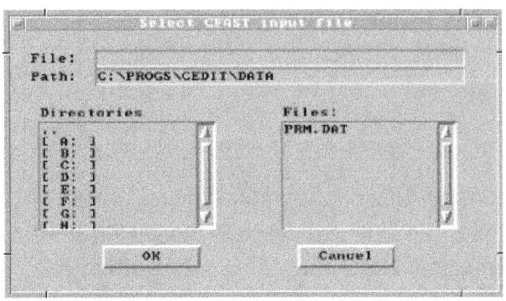

The file specification window consists of four key elements: the filename edit widget, the directory edit widget, the directory selection list, and the filename selection list. To search directories other than the current directory displayed in the directory edit widget, enter the new directory name in the directory edit widget and press Enter. Directories may also be selected by double-clicking on the drive or directory in the directory selection list. If the filename is known, enter it in the filename edit widget and press Enter. Do not include the drive or directory in the filename. Drive and directory specifications are handled only through the directory edit widget and the directory selection list widget. Once the desired file is found, double-click on the entry in the filename selection list, or single-click and press the *OK* button.

Press *Begin* to begin deleting the file, or *Cancel* to exit this procedure without deleting the file.

7 References

[1] Bukowski, R.W., "Strawman Procedure for Assessing Toxic Hazard." In Summary preliminary report of the advisory committee on the toxicity of the products of combustion. NFPA, Quincy, MA; June 1984.

[2] Lawson, J.R. and Quintiere, J.G., "Slide-rule Estimates of Fire Growth," Natl. Bur. Stand. (U.S.), NBSIR 85-3196, Gaithersburg, MD, pp. 56. (1985).

[3] Nelson, H.E., "FIREFORM: A Computerized Collection of Convenient Fire Safety Computations," Natl. Bur. Stand. (U.S.), NBSIR 86-3308 (1986).

[4] Peacock, R. D., Forney, G. P., Reneke, P. A., Portier, R. W., and Jones, W. W., "CFAST, the Consolidated Model of Fire Growth and Smoke Transport," Natl. Inst. Stand. Technol, Technical Note 1299, 241 pp (1993).

[5] Schifiliti, R. P, Meacham, B. J., and Custer, R. L. P., "Design of Detection Systems," Chapter 4-1 in The SFPE Handbook of Fire Protection Engineering, Second Edition, DiNenno, P.J., et. al., editors, National Fire Protection Association, HFPE-95 (1995).

[6] Heskestad, G. and Delichatsios, M.A., "Environments of Fire Detectors - Phase 1: Effect of Fire Size, Ceiling Height, and Material. Volume 2. Analysis," NBS-GCR-77-95, Natl. Bur. Stand. (U.S.), 100pp (1977).

[7] Stroup, D.W. and Evans, D.D., "Use of Computer Models for Analyzing Thermal Detector Spacing," *Fire Safety Journal*. **14**, 33-45 (1988).

[8] "Standard for the Installation of Sprinkler Systems," NFPA 13, 1994 Edition, National Fire Protection Association (1994).

[9] Evans, D. D., "Sprinkler Fire Suppression Algorithm for HAZARD," Natl. Inst. Stand. Technol., NISTIR 5254 (1993).

[10] Drysdale, D., "An Introduction to Fire Dynamics," John Wiley and Sons, New York, 143 p. (1985).

[11] Tewarson, A., "Combustion of Methanol in a Horizontal Pool Configuration," Factory Mutual Research Corp., Norwood, MA, Report No. RC78-TP-55 (1978).

[12] McCaffrey, B. J., "Entrainment and Heat Flux of Buoyant Diffusion Flames," Natl. Bur. Stand. (U.S.), NBSIR 82-2473, 35 p. (1982).

[13] Koseki, H., "Combustion Properties of Large Liquid Pool Fires," Fire Technology, **25(3)**, 241-255 (1989).

[14] Schifiliti, R. P., "Use of Fire Plume Theory in the Design and Analysis of Fire Detector and Sprinkler Response," Master's Thesis, Worcester Polytechnic Institute, Worcester, MA (1986).

[15]   Lawson, J.R., Walton, W.D., and Twilley, W. H., "Fire Performance of Furnishings as Measured in the NBS Furniture Calorimeter," Natl. Bur. Stand. (U.S.), NBS IR 83-2787 (1984).

[16]   Pauls, J., "Movement of People,"Chapter 3-13 in the SFPE Handbook of Fire Protection Engineering, Second Edition, DiNenno, P.J., et. al., editors, National Fire Protection Association, HFPE-95 (1995).

[17]   Nelson, H.E. and McLennan, H.A., "Emergency Movement," Chapter 3-14 in the SFPE Handbook of Fire Protection Engineering, Second Edition, DiNenno, P.J., et. al., editors, National Fire Protection Association, HFPE-95 (1995).

[18]   Evans, D. D. and Stroup, D. W., "Methods to Calculate the Response of Heat and Smoke Detectors Installed Below Large Unobstructed Ceilings," Natl. Bur. Stand. (U.S.), NBSIR 85-3167 (1985).

[19]   Alpert, R. L., "Calculation of Response Time of Ceiling-Mounted Fire Detectors," Fire Technology, Vol 8:(3), National Fire Protection Association, Quincy, MA, pp. 181-195 (1972).

[20]   Evans, D. D., "Calculating Fire Plume Characteristics in a Two-Layer Environment," Fire Technology, Vol. 20:(3), National Fire Protection Association, Quincy, MA, pp. 39-63 (1984).

[21]   Alpert, R.L., and Ward, E. J.; "Evaluating Unsprinklered Fire Hazards;" SFPE Technology Report 83-2; Society of Fire Protection Engineers: Boston, MA (1983).

[22]   Heskestad, G. and Delicatsios, M.A., "Environments of Fire Detectors--Phase I. Effect of Fire Size, Ceiling Height and Material," Natl. Bur. Stand. (U.S.), NBS-GCR-77-86 (1977).

[23]   Walton, W. D., "ASET-B, A Room Fire Program for Personal Computers," Natl. Bur. Stand. (U.S.), NBSIR 85-3144 (1985).

[24]   Zukoski, E. E.; "Development of a Stratified Ceiling Layer in the Early Stages of a Closed-Room Fire," Fire and Materials, Vol. 2, No. 2 (1978).

[25]   Cooper, L. Y. and Stroup, D. W., "Calculating Available Safe Egress Time from Fires," Natl. Bur. Stand. (U.S.), NBSIR 82-2587 (1982).

[26]   Klote, J.H. and Milke, J.A., "Design of Smoke Control Management Systems," ASHRAE and SFPE, Atlanta, GA 30329 (1992).

[27]   Alpert, R.L. and Ward, E. J.; "Evaluating Unsprinklered Fire Hazards," SFPE Technology Report 83-2; Society of Fire Protection Engineers: Boston, MA (1983).

[28]   Zukoski, E.E., Kubota, T. and Cetegen, B., "Entrainment in the Near Field of a Fire Plume. Final Report," Natl. Bur. Stand. (U.S.), NBS-GCR-81-346 (1981).

[29]   Nelson, H.E., "FPETOOL: Fire Protection Engineering Tools for Hazard Estimation," Natl. Inst. Stand. Technol., NISTIR 4380 (1990).

[30] Evans, D., "Ceiling Jet Flows," Chapter 2-4 in the SFPE Handbook of Fire Protection Engineering, Second Edition, DiNenno, P.J., et. al., editors, National Fire Protection Association, HFPE-95 (1995).

[31] Alpert, R.L. and Ward, E. J.; "Evaluating Unsprinklered Fire Hazards;" SFPE Technology Report 83-2, Society of Fire Protection Engineers: Boston, MA (1983).

[32] Heskestad, G. "Fire Plumes," Chapter 2-2 in the SFPE Handbook of Fire Protection Engineering, Second Edition, DiNenno, P.J., et. al., editors, National Fire Protection Association, HFPE-95 (1995).

[33] McCaffrey B.J. "Flame Height," Chapter 2-1 in the SFPE Handbook of Fire Protection Engineering, Second Edition, DiNenno, P.J., et. al., editors, National Fire Protection Association, HFPE-95 (1995).

[34] Quintiere, J.G. and Harkelroad, M.F., "New Concepts for Measuring Flame Spread Properties," Symposium on Application of Fire Science to Fire Engineering: American Society for Testing and Materials and Society of Fire Protection Engineers, Denver, CO, 27 June 1984.

[35] Quintiere, J.G., "A Semi-Quantitative Model for the Burning of Solid Materials," Natl. Inst. Stand. Technol., NISTIR 4840 (1992).

[36] Fire Resistance Tests of Structures, International Organization for Standardization Recommendation R834 (1968).

[37] Law, M., "Prediction of Fire Resistance," Proceedings, Symposium No. 5, Fire-resistance Requirements for Buildings - A New Approach; Joint Fire Research Organization, London, Her Majesty's Stationery Office (1973).

[38] Thomas, P.H. and Heseldon, A.J.M., "Fully-developed fires in single compartments. A co-operative research promgramme of the Conseil International du Bâtiment," Joint Fire Research Organization Fire Research Note No. 923 (1972).

[39] Nelson, H.E., "FPETOOL: Fire Protection Engineering Tools for Hazard Estimation," Natl. Inst. Stand. Technol., NISTIR 4380 (1990).

[40] Lawson, J.R. and Quintiere, J.G., "Slide-rule Estimates of Fire Growth," Natl. Bur. Stand. (U.S.), NBSIR 85-3196 (1985).

[41] Kawagoe, K. D. and Sekine, T., "Estimation of Fire Temperature-Time Curve for Rooms," BRI Occasionial Report No. 11, Building Research Institute, Ministry of Construction, Japanese Government (1963).

[42] McCaffrey, B.J., "Purely Buoyant Diffusion Flames: Some Experimental Results," Combustion Institute: Eastern States Section (1978).

[43] Zukoski E., "Development of a Stratified Ceiling Layer in the Early Stages of a Closed-Room Fire," Fire and Materials, Vol. 2, No. 2 (1978).

[44] Morton, B.R., Taylor, G. and Turner, J.S., "Turbulent Gravitational Convection from Maintained and Instantaneous Sources," Proceedings of the Royal Society of London, Series A, No. 1196, London, England (1956).

[45] Heskestad, G., "Engineering Relations for Fire Plumes," Fire Safety Journal, Vol. 7, pp. 25-32 (1984).

[46] Cooper, L. Y. and Stroup, D. W., "Calculating Available Safe Egress Time from Fires," Natl. Bur. Stand. (U.S.), NBSIR 82-2587 (1982).

[47] Grubitts, S. and Shestopal, V.O., "Computer Program for an Uninhibited Smoke Plume and Associated Computer Software," Fire Technology, Vol 29, (3), pp. 246-67 (1993).

[48] Tewarson, A., "Generation of Heat and Chemical Compounds in Fires," Chapter 3-4 in the SFPE Handbook of Fire Protection Engineering, Second Edition, DiNenno, P.J., et. al., editors, National Fire Protection Association, HFPE-95 (1995).

[49] Babrauskas, V., "Will the Second Item Ignite," Natl. Bur. Stand. (U.S.), NBSIR 81-2271 (1982).

[50] Tewarson, A., "Generation of Heat and Chemical Compounds in Fires," Chapter 3-4 in the SFPE Handbook of Fire Protection Engineering, Second Edition, DiNenno, P.J., et. al., editors, National Fire Protection Association, HFPE-95 (1995).

[51] Klote, J.H. and Milke, J.A., "Design of Smoke Control Management Systems," ASHRAE and SFPE, Atlanta, GA 30329, pp. 21-32 (1992).

[52] Thomas, P. H., "Testing Products and Materials for Their Contribution to Flashover in Rooms," Fire and Materials, 5, pp. 103-111 (1981).

[53] Babrauskas, V., "Upholstered Furniture Room Fires--Measurements, Comparison with Furniture Calorimeter Data, and Flashover Predictions," *Journal of Fire Sciences*, Vol. 2, 5 (1984).

[54] Kawagoe, K. D. and Sekine, T., "Estimation of Fire Temperature-Time Curve for Rooms," BRI Occasionial Report No. 11, Building Research Institute, Ministry of Construction, Japanese Government (1963).

[55] Fang, J.B. and Breese, J.N., "Fire Development in Residential Basement Rooms, Interim Report," Natl. Bur. Stand. (U.S.), NBSIR 80-2120 (1980).

[56] Babrauskas, V., "COMPF2--A Program for Calculating Post-Flashover Fire Temperatures," Natl. Bur. Stand. (U.S.), Technical Note 991 (1979).

www.ingramcontent.com/pod-product-compliance
Lightning Source LLC
Chambersburg PA
CBHW081830170526
45167CB00007B/2776